ENHANCEMENT OF BIOLOGICAL NITROGEN FIXATION
OF COMMON BEAN IN LATIN AMERICA

Developments in Plant and Soil Sciences

VOLUME 52

Enhancement of Biological Nitrogen Fixation of Common Bean in Latin America

Results from an FAO/IAEA Co-ordinated Research Programme, 1986-1991

Implemented by
the Soil Fertility, Irrigation and Crop Production Section, Joint FAO/IAEA Division
and
the Soil Science Unit, FAO/IAEA Agriculture Laboratory

Edited by

F.A. BLISS
Department of Pomology,
University of California at Davis,
Davis, CA 95616, USA

G. HARDARSON
Joint FAO/IAEA Programme,
Soil Science Unit,
Agency's Laboratories,
A-2444 Seibersdorf, Austria

Reprinted from *Plant and Soil,* Volume 152, No. 1 (1993)

KLUWER ACADEMIC PUBLISHERS
Dordrecht / Boston / London

in cooperation with
FOOD AND AGRICULTURE ORGANIZATION
of the United Nations and the
INTERNATIONAL ATOMIC ENERGY AGENCY
Joint FAO/IAEA Division - IAEA Laboratories

Library of Congress Cataloging-in-Publication Data

Enhancement of biological nitrogen fixation of common bean in Latin America : results from an FAO/IAEA co-ordinated research programme, 1985-1991 / edited by F.A. Bliss, G. Hardarson.
p. cm. -- (Developments in plant and soil sciences ; v. 52)
Includes bibliographical references.
ISBN 0-7923-2451-X (HB : acid free paper)
1. Kidney bean--Latin America. 2. Nitrogen-fixing plants--Latin America. 3. Nitrogen--Fixation. I. Bliss, Fredrick A., 1938- . II. Hardarson, G. (Gudni) III. Food and Agriculture Organization of the United Nations. IV. International Atomic Energy Agency. V. Series.
SB327.E53 1993
633.3'72--dc20 93-26850

ISBN 0-7923-2451-X

Published by Kluwer Academic Publishers,
P.O. Box 17, 3300 AA Dordrecht, The Netherlands.

Kluwer Academic Publishers incorporates
the publishing programmes of
D. Reidel, Martinus Nijhoff, Dr W. Junk and MTP Press.

Sold and distributed in the U.S.A. and Canada
by Kluwer Academic Publishers,
101 Philip Drive, Norwell, MA 02061, U.S.A.

In all other countries, sold and distributed
by Kluwer Academic Publishers Group,
P.O. Box 322, 3300 AH Dordrecht, The Netherlands.

Printed on acid-free paper

Printed in the Netherlands

Contents

Foreword

Legumes are cultivated in climates ranging from tropical through sub-arctic and arid to humid. They are particularly important due to their ability to fix atmospheric nitrogen in symbiosis with root-nodule *Rhizobium* bacteria and due to their high nutrient value for human food and animal feed. In many developing countries legumes are the main source of protein and an essential dietary component. In spite of their importance, legumes often have low yields and low unstable harvests. The plants are prone to lodging and flower shedding, and the pods to shattering. Much too little efforts have been invested in improving leguminous crops by breeding improved cultivars and most of this effort has been concentrated on cash crops such as soybean and groundnuts.

Common bean (*Phaseolus vulgaris* L.), which is also known in English as Kidney, haricot, French or snap bean, originated from Central and South America. It is the most important pulse crop in that area, as well as in large parts of Asia and Africa.

Biological nitrogen fixation in leguminous crop has been studied in several research programmes supported and co-ordinated by the Soil Fertility, Irrigation and Crop Production Section of the Joint FAO/IAEA Division and the FAO/IAEA Agriculture Laboratory in Seibersdorf, Austria. Research, performed prior to that reported in this book, had demonstrated that great differences exist between legume species in their abilities to support biological nitrogen fixation. In these studies the common bean was found to be rather poor in fixing atmospheric nitrogen. An FAO/IAEA Consultants Meeting held in Vienna in September 1983 recommended the initiation of a Co-ordinated Research Programme to address this problem. The proceedings of this meeting were published in 1984,* including general recommendations on breeding for improved nitrogen fixation in grain legumes.

The FAO/IAEA Co-ordinated Research Programme reported in this book had participants from Brazil, Chile, Colombia, Guatemala, Mexico, Peru and the United States and was conducted during the period 1986–1991. The programme to 'Enhance N_2 Fixation in Common Bean' was co-ordinated by the Soil Fertility, Irrigation and Crop Production Section of the Joint FAO/IAEA Division which supported the programme by awarding research grants, and conducting research co-ordination meetings. Isotope analytical services were performed by the FAO/IAEA Agriculture Laboratory.

The information contained in this book will hopefully help scientists working on biological nitrogen fixation and legume production in enhancing N_2 fixation in this important crop and therefore in increasing food production especially in developing countries without resorting to increased nitrogen fertilizer inputs.

Björn Sigurbjörnsson
Director
Joint FAO/IAEA Division
Vienna, Austria

* Hardarson G. and Lie T A 1984 *Breeding Legumes for Enhanced Symbiotic Nitrogen Fixation* (Dordrecht: Martinus Nijhoff/Dr. W. Junk Publishers, Kluwer Academic Publishers Group, First published in *Plant and Soil*, 82 (3); Advances in Agricultural Biology, Vol. 12).

Editorial note

List of Participants in FAO/IAEA Programme Enhancement of Biological Nitrogen Fixation of Common Bean in Latin America

Consultants' Meeting for the planning of programme (1983)

C.A. Atkins, The University of Western Australia
D.K. Barnes, University of Minnesota, USA
E. Jacobsen, University of Groningen, The Netherlands
T.A. Lie, Agricultural University, Wageningen, The Netherlands
L.R. Mytton, Welsh Plant Breeding Station, UK
P.S. Nutman, Exeter, UK
R. Sylvester-Bradley, CIAT, Colombia

(The Proceedings of the FAO/IAEA Consultants' Meeting held in 1983 were published in *Plant and Soil*, Vol. 82, No. 3 (1984), and in *Breeding Legumes for Enhanced Symbiotic Nitrogen Fixation*, ed. by G. Hardarson and T.A. Lie, 1984. Dordrecht: Martinus Nijhoff/Dr. W. Junk Publishers.)

FAO/IAEA Co-ordinated Research Programme to enhance biological nitrogen fixation of common bean in Latin America (1986–1991)

Agreement holders:

D.K. Barnes, University of Minnesota, USA*
F.A. Bliss, University of California, Davis, USA
J.A. Kipe-Nolt, CIAT, Colombia

Contractors:

R. Aguilera, DGEN, Guatemala*
M.R. Cigales-Rivero, University of Colima, Mexico
R.A. Henson, EMBRAPA-CNPAF, Brazil
L. Longeri, University of Concepcion, Chile
A. Manrique, University of La Molina, Peru
J.J. Peña-Cabriales, CINVESTAV, Mexico
P.A.A. Pereira, EMBRAPA-CNPAF, Brazil
S.M. Tsai, CENA, Brazil
C.A. Sanabria, DGEN, Guatemala
C. Villagran, DGEN, Guatemala*

Co-ordination:

G. Hardarson, FAO/IAEA, Austria

* Participated in the initial phase of the programme.

Acknowledgements

Thanks are due to the staff of the Soil Fertility, Irrigation and Crop Production Section of the Joint FAO/IAEA Division in Vienna, the Soil Science Unit of the FAO/IAEA Agricultural Laboratory, Seibersdorf, the IAEA Research Contract Section and the IAEA Publishing Section for extensive help during the FAO/IAEA Co-ordinated Research Programme on common bean in Latin America 1986–1991 and during the publication of the present issue. We are grateful to Ms H. Axmann and the analytical group at the IAEA Laboratory for isotope analyses for all participants in the programme and to Ms R.C. Rossi for typing the manuscripts.

Plant and Soil **152**: 1–17, 1993.

PLSO BNF01

Methods for enhancing symbiotic nitrogen fixation

GUDNI HARDARSON
FAO/IAEA Programme, Soil Science Unit, Agency's Laboratories, A-2444 Seibersdorf, Austria

Key words: grain legumes, N_2, ^{15}N isotope dilution, nitrogen fixation, *Rhizobium*, symbiosis

Abstract

Biological nitrogen fixation of leguminous crops is becoming increasingly important in attempts to develop sustainable agricultural production. However, these crops are quite variable in their effectiveness in fixing nitrogen. By the use of the ^{15}N isotope dilution method some species have been found to fix large proportions of their nitrogen, while others like common bean have been considered rather inefficient. Methods for increasing N_2 fixation are therefore of great importance in any legume work. Attempts to enhance nitrogen fixation of grain legumes has been mainly the domain of microbiologists who have selected rhizobial strains with superior effectiveness or competitive ability. Few projects have focused on the plant symbiont with the objective of improving N_2 fixation as done in the FAO/IAEA Co-ordinated Research Programme which is being reported in this volume. The objective of the present paper is to discuss some possibilities available for scientists interested in enhancing symbiotic nitrogen fixation in grain legumes. Examples will be presented on work performed using agronomic methods, as well as work on the plant and microbial symbionts. There are several methods available to scientists working on enhancement of N_2 fixation. No one approach is better than the others; rather work on the legume/*Rhizobium* symbiosis combining experience from various disciplines in inter-disciplinary research programmes should be pursued.

Introduction

Nitrogen, an element vital for plant growth constitutes 78% of the earth's atmosphere. In spite of its abundance it is one of the most limiting factors for crop growth, and nitrogen fertilizers represent one of the major costs in crop production.

Nitrogen can also be supplied to crops by biological nitrogen fixation, a process which is becoming more important for not only reducing energy costs, but also attempting to develop sustainable agricultural production. Legumes in symbiosis with rhizobia, are the best known nitrogen fixing systems. To reduce nitrogen fertilizer use and potential ground water pollution it will be necessary to increase the use of nitrogen fixing systems to maintain or increase crop yields. Nitrogen fixing micro-organisms will therefore be an important component of sustainable agricultural systems.

Tauer (1989) recently stated that 'increasing the efficiency of legumes to fix N_2 may have an annual US benefit of $1,067 million while decreasing N fertilization by 1,547 thousand metric tons. Total elimination of N fertilization of the major crops has an annual US benefit of $4,484 million'. Economic evaluations of the benefit of biological nitrogen fixation have not been performed for developing countries but it is likely that comparable benefits could be expected.

Biological nitrogen fixation is important from the point of view of saving of N fertilizer and reducing the cost of crop production. Also, the use of nitrogen fixing crops will (i) reduce ground water pollution in comparison to cereal crops, which have to be fertilized; (ii) enhance protein production due to the high protein content of legume crops; (iii) contribute N for succeeding crops and (iv) build up soil fertility. Some of these factors are particularly important in developing countries, others in developed ones. It is therefore urgent to promote further use of the legume/*Rhizobium* symbiosis.

Several excellent books have been written on the legume/*Rhizobium* technology (Somasegaran and Hoben, 1985; Vincent, 1970); on legume inoculation (FAO, 1984) and on methods to quantify biological nitrogen fixation (Bergersen, 1980, Peoples et al., 1989). Only a few have dealt with breeding legumes for enhanced nitrogen fixation (Hardarson and Lie, 1984, Viands et al., 1981) and none, as far as I know, has been written exclusively on methods for enhancing symbiotic nitrogen fixation, although it is the ultimate aim of most legume/*Rhizobium* research. It is quite disturbing that the relatively simple *Rhizobium* inoculant technology has not been transferred to many of the developing countries. At the same time some institutes are starting to use modern biotechnology and molecular biology with the objective of enhancing N_2 fixation. The gap between countries in terms of technology used is widening. Modern biotechnology holds some promise as a tool for increasing nitrogen fixation but there are many other promising methods available, which have not been investigated fully.

The objective of this present paper is to discuss the possibilities for enhancing N_2 fixation by working on the plant and microbial symbionts and by using agronomic methods. Examples will be taken from research work performed at the FAO/IAEA laboratory, from experiments conducted as part of FAO/IAEA Co-ordinated Research Programmes or elsewhere from the literature.

Effectiveness of the legume/*Rhizobium* symbiosis

The effectiveness of different legume species and their microsymbionts have been listed in several publications (Hardarson et al., 1987, FAO, 1984). As an example faba bean (*Vicia faba*), *lupin* (*Lupinus* spp.), and pigeon peas (*Cajanus cajan*) have been found to be very efficient; soybean (*Glycine max*), groundnut (*Arachis hypogaea*) and cowpea (*Vigna unguiculata*) to be average; and common bean (*Phaseolus vulgaris*) and pea (*Pisum sativum*) rather poor in fixing atmospheric nitrogen. Some examples of % N derived from atmosphere (% Ndfa) in various grain legumes as measured by ^{15}N isotope dilution are presented in Fig. 1. The average values % Ndfa for several grain legumes as obtained by investigators participating in an FAO/IAEA Co-ordinated Research Programme (Danso, unpublished) are shown. There were great differences among the grain legume species, with faba bean

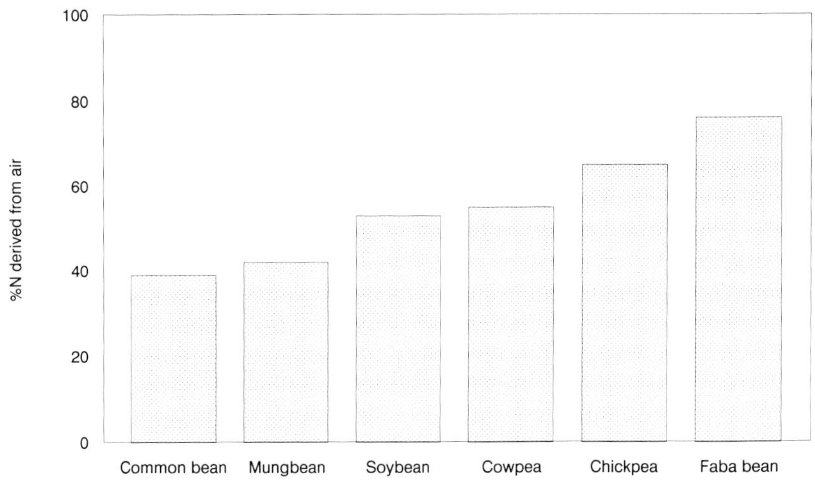

Fig. 1. Percentage N derived from atmosphere (%Ndfa) of several grain legume species. The results are average values of measurements made in an FAO/IAEA Co-ordinated Research Programme with participants from several countries (Danso, unpublished data).

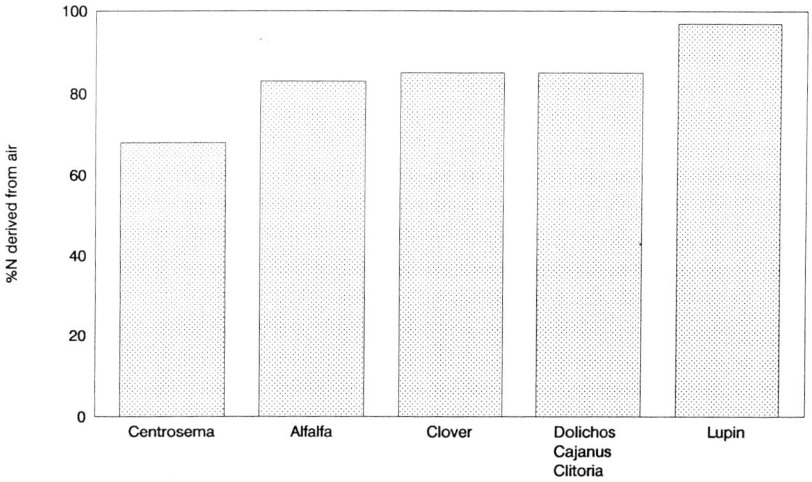

Fig. 2. Percentage N derived from atmosphere of several forage legumes. The results are average values of measurements made in an FAO/IAEA Co-ordinated Research Programme with participants from several countries (Bowen and Danso, 1987)

being very effective, common bean rather poor and others intermediate. It is clear from the above that it should be possible to enhance N_2 fixation in some of these species especially those that derive low or average amounts of N from the atmosphere. It will be more difficult to increase nitrogen fixation in those plants that already derive a high proportion of total nitrogen from the atmosphere, i.e., faba bean.

Forage or pasture legumes are usually more efficient in nitrogen fixation and derive high percentages of their N from the atmosphere (Fig. 2, Bowen and Danso, 1987). The figure illustrates average % Ndfa for number of forage legume species as measured by the isotope dilution method in number of countries. All the forage legumes fixed a high proportion of their nitrogen from the atmosphere, contrary to the performance of the grain legumes. However they were quite variable in terms of yield, which will affect the amount of N_2 fixed.

Factors, affecting symbiotic nitrogen fixation

Agronomic and environmental factors

Environmental factors affecting nitrogen fixation include temperature, moisture, acidity and several chemical components of the soil such as nitrogen, phosphorus, calcium and molybdenum content (FAO, 1984). It is often difficult to isolate the effect of the above factors on inoculation success from their influence on symbiosis and nitrogen fixation. For example, acidity, as well as, calcium, aluminium and manganese concentration will interact and affect both bacterial proliferation, root-hair infection and plant growth (FAO, 1984). Temperature and moisture may also affect inoculation success with survival of rhizobia being affected by high temperature, although large differences in tolerance to high temperature have been reported (IITA, 1985). Low temperature may also affect symbiosis and nitrogen fixation (Hardarson and Jones, 1979a). Poor survival of *Rhizobium* has been reported in flooded soil, but much less is known about the effect of soil dehydration on inoculation success (Roughley, 1980). Few recommendations have been proposed for methods to ameliorate the adverse temperature or moisture effects, except that surface mulches may conserve moisture and reduce soil temperature (Roughly, 1980).

Generally combined N delays or inhibits nodulation and nitrogen fixation (Fig. 3). Because of this adverse effect, N fertilization usually is not recommended for leguminous crops. However there may be situations where N fertilizers have to be applied such as to cereals in mixed cropping or rotations. There, fertilizer may affect N_2 fixation of the legume crops.

Differences have been found among soybean

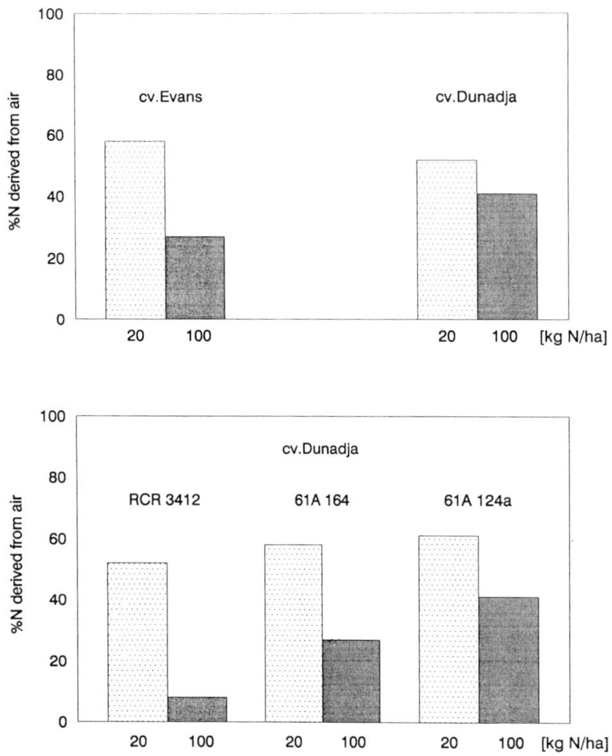

Fig. 3. Effect of combined nitrogen on N_2 fixation of soybean/*Bradyrhizobium* symbiosis (Hardarson et al., 1984; Senaratne et al., 1987).

cultivars and rhizobial strains for tolerance to N fertilization (Fig. 3). In a field experiment 20 and 100 kg N ha^{-1} were applied to various cultivars and nitrogen fixation measured by the ^{15}N isotope dilution method (Hardarson et al., 1984b). Differences were found between cultivars for tolerance to N fertilizer application with some (cv. Evans) being rather sensitive and others (cv. Dunadja) somewhat tolerant. However, later greenhouse experiments revealed that this could be affected by rhizobial inoculation (Senaratne et al., 1987).

It is possible to apply small amounts of soil or foliar N fertilizer, which may increase yield without reducing the amount of nitrogen fixed (Fig. 4). This field experiment was performed by applying 20 kg N ha^{-1} as starter and additional 40 kg N ha^{-1} as soil and/or foliar fertilizer to soybean (Afza et al., 1987). Additional 40 kg N ha^{-1} soil or foliar fertilizer application increased the seed yield significantly without reducing the amount of N_2 fixed. However, additional fertilization with 40 + 40 kg N ha^{-1} reduced nitrogen fixation. Other studies have reported less effect of foliar fertilization (Boote et al., 1978; Poole et al., 1983).

An interesting characteristic of faba bean is its tolerance to combined N (Fig. 5). The field experiments illustrated in this figure were performed during two years in the field of Seibersdorf laboratory (Hardarson et al., 1991). Different rates of N fertilizer were applied to the faba bean and a non-fixing control, with nitrogen fixation measured by the ^{15}N isotope dilution method. In the above study and that of Roughley et al. (1983) faba beans were able to fix a substantial percentage and amount of N even when 200 and 400 kg N ha^{-1} fertilizer were applied. Faba bean would therefore be particularly suitable for rotation or mixed cropping situations where N fertilizer has to be applied to the companion cereal crop.

The compatibility of rhizobia with pesticides is poorly understood except for fungicides. Chemical dressings with heavy metals such as mercury, zinc, copper or lead are toxic to rhizobia. Organic fungicides, e.g. Captan, Carboxin, Chloranil, PCNB or Thiram, are slightly less toxic (FAO, 1984). Other fungicides have been found to have a variable effect on different rhizobial strains (Ramos, et al., 1993). However, insecticides have little adverse effect on nodulation when not directly applied on seed. The effect of herbicides on rhizobial survival and nodulation is unknown.

Effect of the micro-symbiont

The rhizobia involved in nodulation can influence the percentage and amount of nitrogen fixed by the legume/*Rhizobium* symbiosis. There are many reports on the effect of rhizobial strains on nitrogen fixation. The effectiveness of 20 strains of *Bradyrhizobium japonicum* in symbiosis with one cultivar of soybean is shown in Figures 6 and 7 (Hardarson et al., 1984a). Nitrogen fixation varied from 38 to 76% as quantified by the ^{15}N isotope dilution method under field condition at the Seibersdorf laboratory (Fig. 6). Amount of N_2 fixed was 60 kg N ha^{-1} for the poorest strain and about 200 kg N ha^{-1} for the most effective one (Fig. 7).

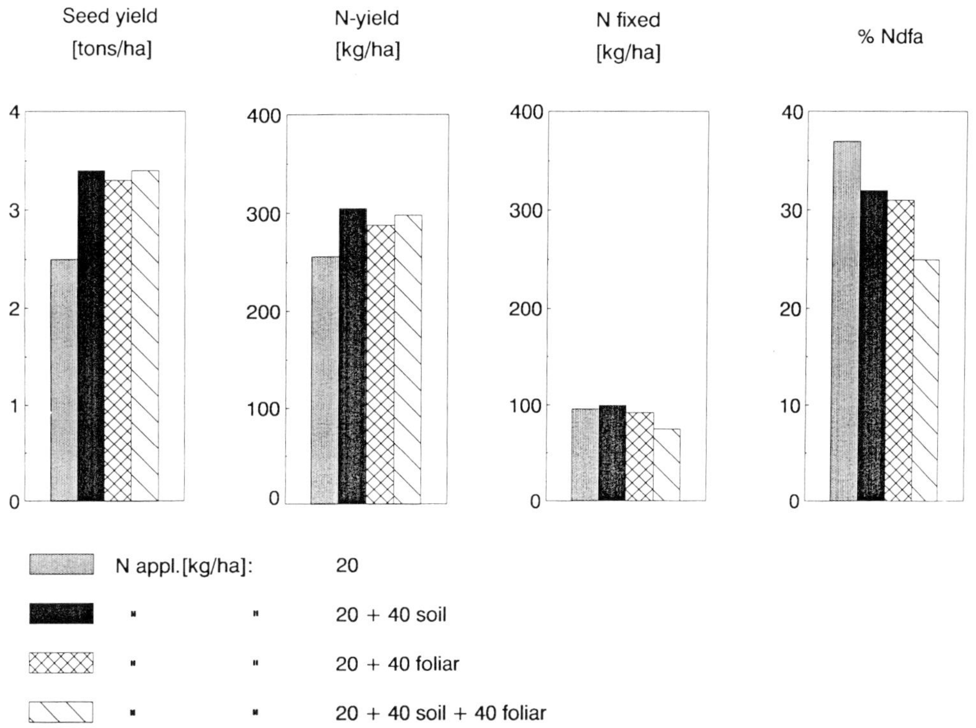

Fig. 4. Effect of delayed soil and foliar N-fertilization on yield and N_2 fixation of soybean (Afza et al., 1987).

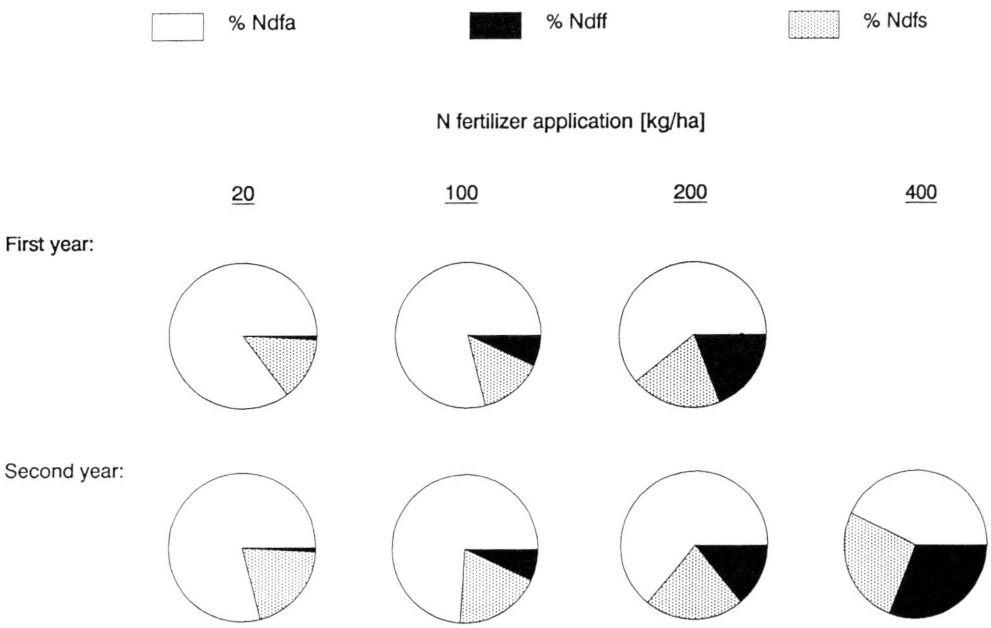

Fig. 5. Percentage N derived from atmosphere (%Ndfa), soil (%Ndfs) and fertiliser (%Ndff) of fababean as affected by various N fertilizer rates (Hardarson et al., 1991).

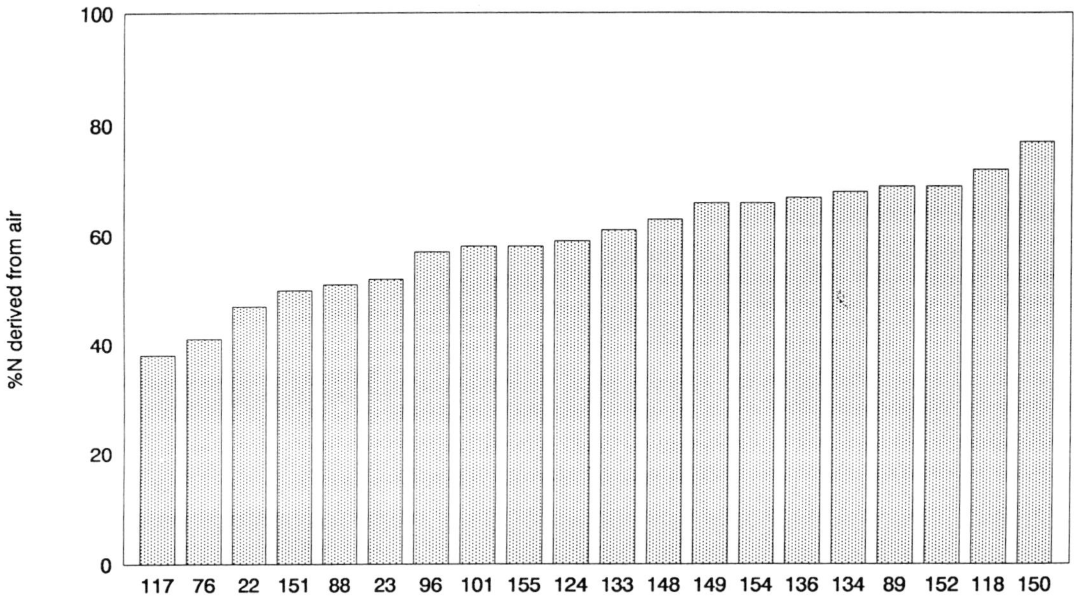

Fig. 6. Field evaluation of N_2 fixation (%Ndfa) by strains of *Bradyrhizobium japonicum* on cv. Chippewa soybean. The strains (61A76 etc) were all obtained from the Nitragin Co. (Presently LiphaTech, Milwaukee, USA) (Hardarson et al., 1984)

No significant differences were observed in the amount of dry matter resulting from the different inoculation treatments, since N was not a limiting factor in the soil. These results illustrate also the advantages of the isotope method for quantifying nitrogen fixation even when no dry matter yield differences were observed.

The proportion of the nodules formed on a particular host is influenced by the competitive ability of an inoculated *Rhizobium* strain in comparison to indigenous strains, which may vary in their effectiveness. The different competitive abilities of two effective strains (102F51 and 102F34) against one ineffective strain (102F62) for nodulating alfalfa (*Medicago sativa*) as well as the variation between alfalfa varieties in their preference for rhizobial strains are shown in Fig. 8. Strain 102F51 was significantly ($p < 0.05$) more competitive than 102F34 in nodule formation on both cultivars.

Other important characteristics of rhizobial inoculants are their ability to survive in the inoculant and in soil, to colonize the rhizosphere and to migrate in soil. It is hoped that further research on rhizobial ecology will illustrate the relationship between population size of rhizobia in the rhizosphere and nodulation.

Effect of the host plant

The host plant cultivars vary greatly in terms of their nitrogen fixation supportive traits and amount of N_2 fixed. This is illustrated by Hardarson et al. (1993).

Methods for enhancing symbiotic nitrogen fixation

Agronomic methods

Agronomic factors can be manipulated with the objective of enhancing symbiotic nitrogen fixation. The effect of limiting factors should be reduced by keeping moisture, acidity, phosphorus, nitrogen, calcium and molybdenum optimal. The effect of pesticides on N_2 fixation should be minimized by separate placement of rhizobia and pesticides.

Methods of rhizobial inoculation can have great influence on the amount of N_2 fixed. One example is shown in Fig. 9, where seed and soil inoculation produced very different nodulation patterns (Hardarson et al., 1989). Previously, the N_2 fixation potential of nodules on lateral

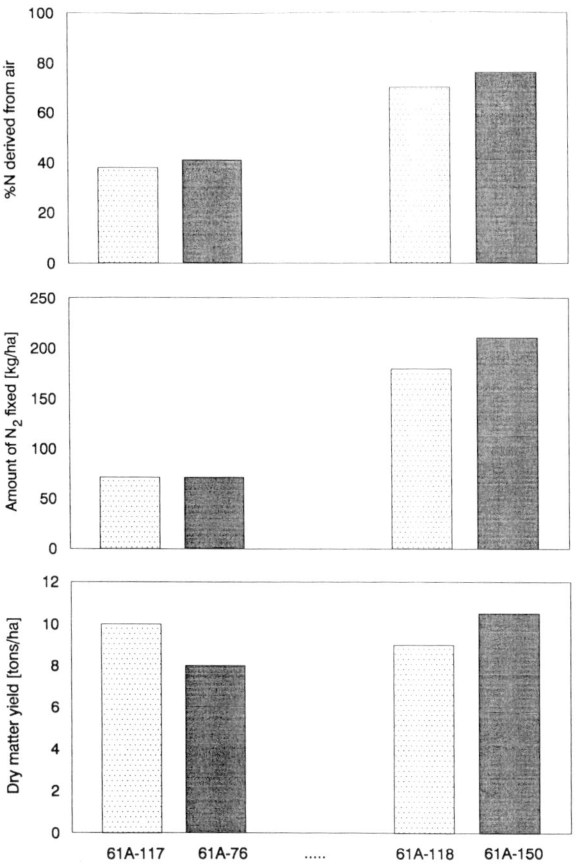

Fig. 7. Same data as in Figure 6 also presenting amount of N_2 fixed and dry matter yield produced by the poorest and the best *Bradyrhizobium* strains.

roots and on the lower part of the root system has been largely discounted and they have been considered much less important than the nodules formed on the crown or on the top most part of the root system.

Nodules on the lower part of the root system can fix more nitrogen over the whole growing season than the crown nodules, and they may contribute most of the nitrogen fixed by the legume plant. The response was clearer in the greenhouse experiment than in the field due to greater contrast in nodulation patterns (Figs. 9 and 10). The quantification of N_2 fixation was performed by the use of ^{15}N isotope dilution method.

Differences in the location of nodules on roots were induced through varying the bacterial inoculation procedure, timing or depth of inoculum placement in the soil. Movement of rhizobia in the soil and rhizosphere was very limited, and nodule formation was restricted primarily to the vicinity of the bacterial placement. Other investigators also have observed limited movement of rhizobia in soil and rhizosphere (McDermott and Graham, 1989) and substantial N_2 fixation by nodules on lateral roots (Wolyn et al., 1989). This could be one reason why only a small proportion of the soybean nodules are usually formed by the inoculated strains (Kuykendall and Weber, 1978). Farmers applying inoculum on the seed can therefore not expect these bacteria to form nodules on the whole root system. It is likely that applied rhizobia form some or most of the nodules on the crown but others indigenous to the soil form the nodules at greater depth and distances from the crown. Thus the effectiveness of these indigenous bacteria greatly affect the amount fixed N_2 by the plant even when inoculant was applied to the seed or into the seed furrow.

In another study (Zapata et al., 1987), the time course of nitrogen fixation of soybean was investigated (Figs. 11 and 12). That field experiment showed low initial nitrogen fixation by soybean. Only 74 days after planting at the beginning of reproductive stage (R_1) was N_2 fixation very effective ($3\,kg\,N\,ha^{-1}\,day^{-1}$); but this high efficiency continued for only 20 days. During that time the root system is well developed and nodules have been formed by indigenous bacteria on the lateral roots.

It should be possible to enhance N_2 fixation of soybean at least 10–20% by optimal production of nodules on lateral roots. After appropriate testing under several environmental conditions inoculation methods could be improved to facilitate the inoculation of the whole root system of the host plant with effective rhizobial strains, and not only the crown or part of the root system, which is done when the seed is inoculated or granular inoculant is applied in the row. Another possibility would be to select rhizobia not only for effectiveness but also for migration in soil and along the root under a range of conditions.

There are several other considerations to be taken into account when optimizating inocula-

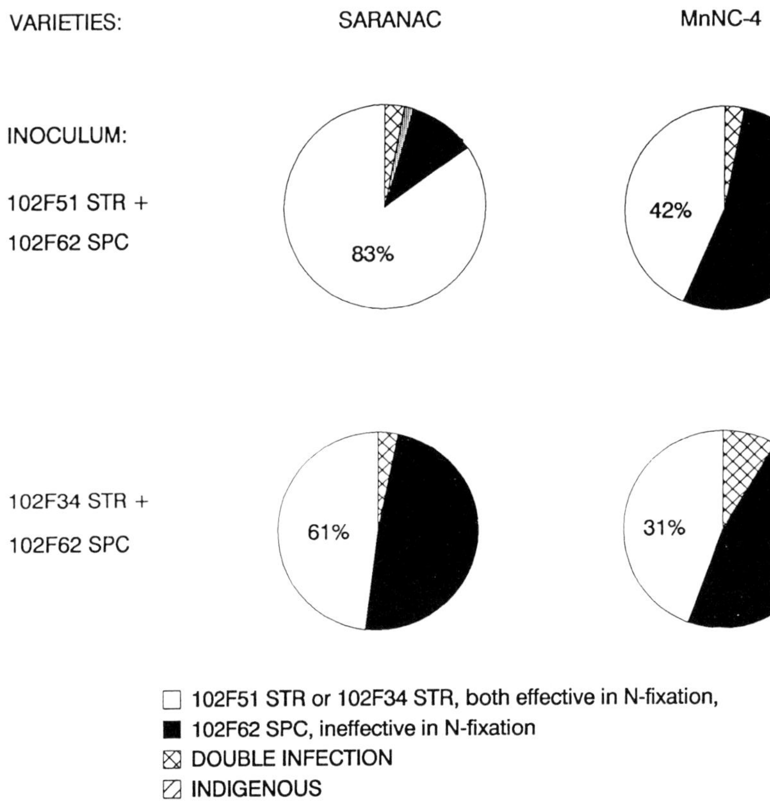

Fig. 8. Proportion of nodules produced by effective and ineffective strains of *Rhizobium melitoti* on alfalfa (Hardarson et al., 1981).

tion methods and these have been reviewed recently (FAO, 1984).

Intercropping legumes with non-legumes is a common cropping system in the tropics. Several reports have demonstrated that the growth of a non-legume can be increased by the presence of the legume (Agboola and Fayemi, 1972; Eaglesham et al., 1981). The non-legume also can affect the effectiveness of N_2 fixation of the legume crop (Fig. 13). In a field experiment (Danso et al., 1987) N_2 fixation in faba bean was affected by plant population density in sole or intercropped systems with barley. In the mixed culture up to 96% of the N in the faba bean crop was derived from the air as compared to 65% when no such competition occurred. Therefore, it seems that legumes grown in a mixed culture must rely more on atmospheric nitrogen than on soil or fertilizer N than when it is grown in a pure stand. Similarly, enhanced N_2 fixation in alfalfa as influenced by the growth in a mixed stand with ryegrass was later observed (Hardarson et al., 1988).

Microbiological methods

(*a*) *Screening of* Rhizobium

Rhizobial inoculant for a particular legume species can be obtained either from other research laboratories, or through a selection programme. Selection of rhizobia is justified only when no suitable strain is available from other sources. Strain selection will be required, for example: (i) when the legume of interest is an uncommon species for which there is no recommended strain and (ii) when inoculation with a recommended strain does not produce adequate nodulation and fixation (Halliday, 1984). If strain selection must be performed, at least 50–100 rhizobial isolates from the same

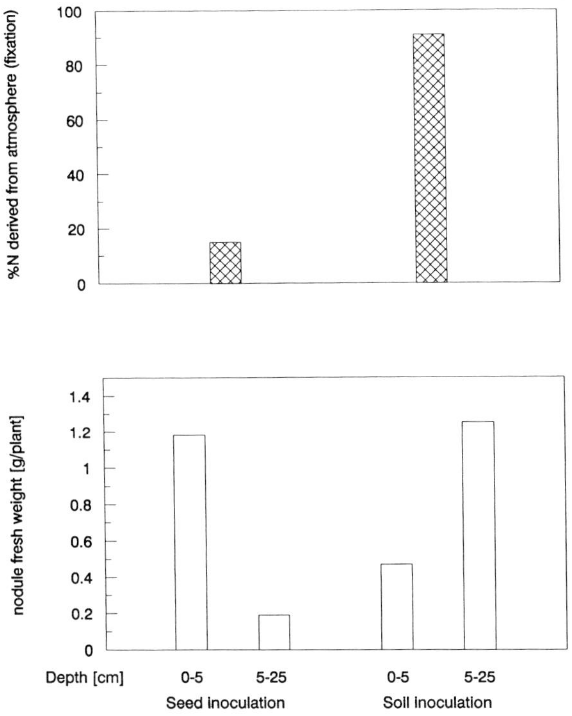

Fig. 9. Nitrogen fixation of soybean as affected by nodulation patterns. Greenhouse experiment. (Hardarson et al., 1989).

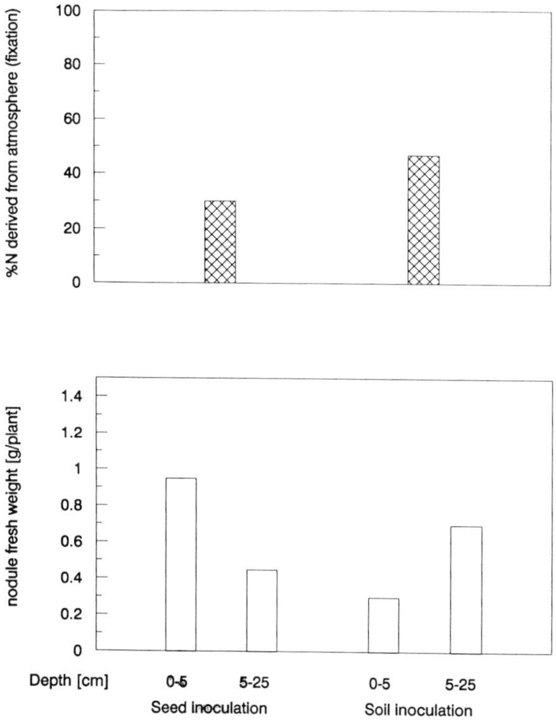

Fig. 10. Nitrogen fixation of soybean as affected by nodulation patterns. Field experiment. (Hardarson et al., 1989).

species or others within the same cross-inoculation group should be included. Preferably, these should be from diverse geographic regions.

The screening of rhizobial isolates should be done in several stages. The first step yields data on infectivity or nodule production under aseptic conditions. Various test assemblies (Gibson, 1980) such as plugged tubes containing agar, semi-enclosed tubes, plastic pouches or Leonard jar can be used. Tests should be performed in either a growth cabinet or greenhouse. The objective is to eliminate before the next step all isolates that do not nodulate the legume under study (Vincent, 1970). At the time of harvest it is usually adequate to observe nodulation and the presence or absence of leghemoglobin.

After the initial infection test, rhizobia should be screened for effective N_2 fixation. This can be done in the same type of test assembly as the infection test or even using open pots, with N deficient medium if sterilized prior to use and care is taken that water is not splashed between pots during watering. Also, if open pots are used, it is important to inoculate with a high population of rhizobia. The probability of a nodule being formed by a strain from another pot is then very low. At the time of evaluation it is usually adequate to measure plant and nodule dry weight and nodule numbers. Plant dry weight is usually well correlated to effectiveness in N_2 fixation, when N is the only limiting growth factor. This test should reduce the number of strains to approximately ten.

The next step should be to evaluate the strains for effective N_2 fixation under soil conditions and at the same time the competitive ability of these strains against indigenous soil bacteria. This test can be performed in the greenhouse using pots and the soil type for which the inoculum will be used. Another possibility is to use soil cores as described by Silvester-Bradley et al. (1983). Results obtained by soil cores can be very different from evaluation made with disturbed soil. During this step, the number of promising strains should be reduced even further.

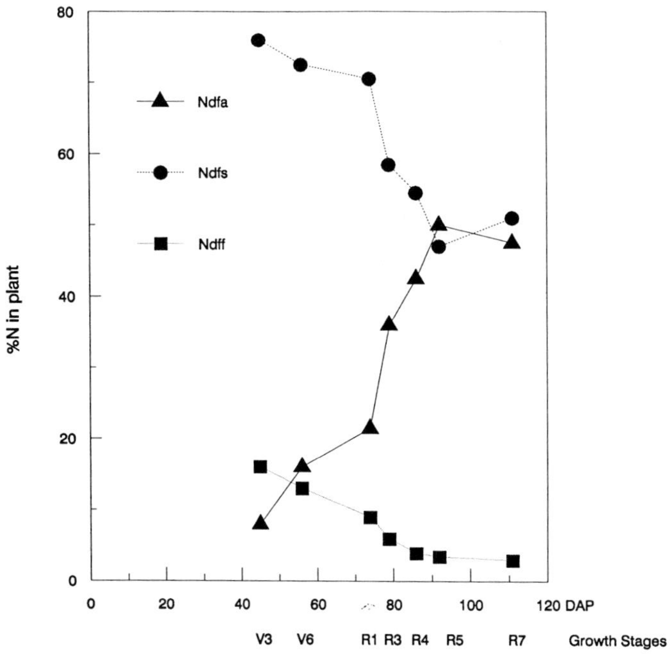

Fig. 11. Proportion of N derived from ^{15}N labelled fertilizer (%Ndff), soil (%Ndfs) and air (%Ndfa) during the growth cycle of Chippewa soybean (Zapata et al., 1987).

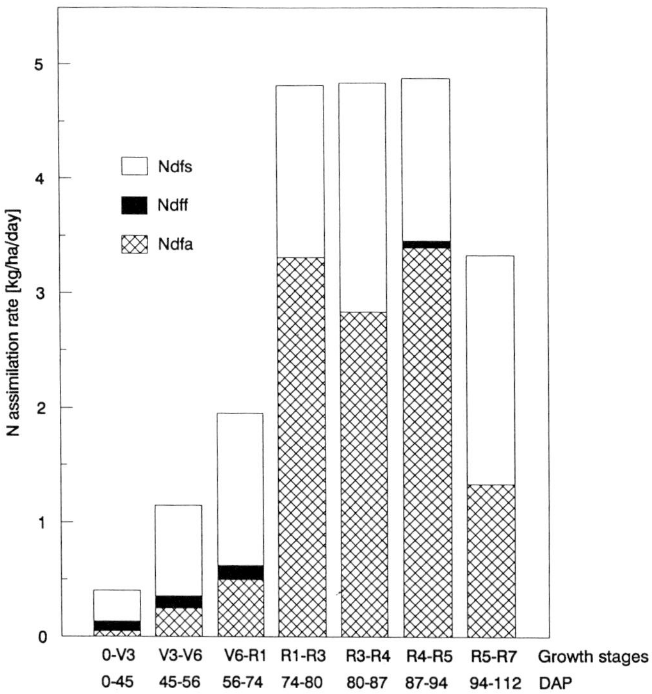

Fig. 12. N assimilation rates of fertilizer (Ndff), soil (Ndfs) and N fixed (Ndfa) by Chippewa soybean at various growth intervals (Zapata et al., 1987).

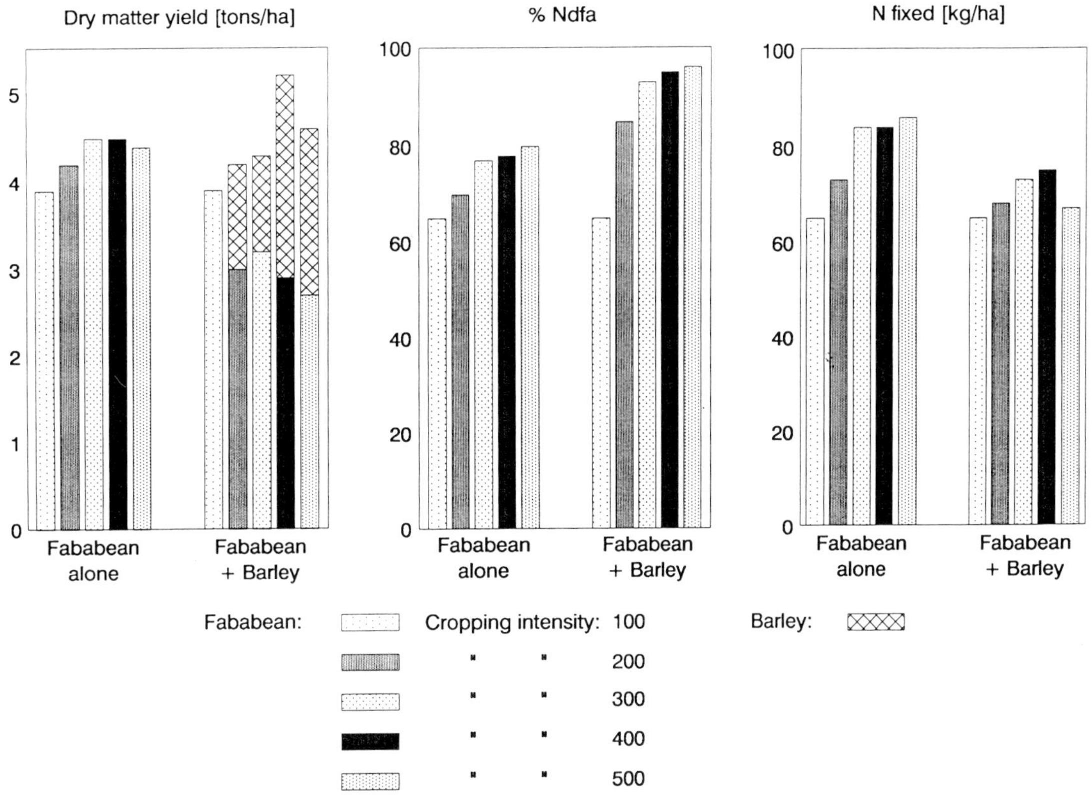

Fig. 13. N_2 fixation by fababean as affected by plant population density in sole or intercropped systems with barley (Danso et al., 1987).

The next step in screening rhizobial strains is to perform field evaluation; first single and later multi-location testing. These experiments should include only three to five of the best strains from the previous steps, since it is usually too expensive to conduct field experiments with numerous strains. Field testing should include evaluation of parameters such as nodulation, yield, nitrogen content, nodule occupancy by rhizobial strains and evaluation of nitrogen fixation, for example, by the use of ^{15}N isotope dilution method. It is preferable to use several parameters when the strains are being ranked for symbiotic performance (Hardarson et al., 1984a).

A stepwise screening procedure should yield highly effective and possibly competitive rhizobial strains. Further screening is needed to ensure high competitive ability against a range of indigenous rhizobia. Survival and persistence should also be ensured. The main criteria for good inoculant strains include (Brockwell et al., 1982):

1) Effective N_2 fixation over a range of environmental conditions.
2) Competitive ability against other strains.
3) Ability to multiply in broth and survive in peat.
4) Survival in seed pellets.
5) Persistence in soil.
6) Ability to migrate in soil and colonize it.
7) Ability to survive adverse environmental conditions.
8) Strain stability during storage.
9) Nodule formation and N_2 fixation in the presence of soil nitrogen.

(b) *Molecular biology methods*

Recent molecular biology methods have made it possible to e.g., (i) produce a genetic map of *Rhizobium*; (ii) develop genetically engineered rhizobia, and (iii) identify strains of *Rhizobium* by DNA probes or marker genes.

The mapping of genes involved in N_2 fixation and nodulation specificity is being performed by several laboratories. It is a very important activity paving the way for genetically engineered rhizobia to be produced. At the same time strains being unusually effective must be identified. The same applies to strains having superior competitive ability and other characters important for successful nodulation and nitrogen fixation. Only after such superior strains have been identified as well as the genes responsible for those traits can one expect to enhance the symbiotic characteristics by transferring these genes to other recipient strains.

A recent article by Craswell (1990) has indicated the difficulties in producing genetically engineered rhizobia with enhanced nitrogen fixation. He writes: 'Modern biotechnology holds much promise as a tool for enhancing biological nitrogen fixation. New technologies for biological nitrogen fixation based on research on genetic engineering have, however, been slow to appear despite considerable investment. Beringer (1989) believes that too little is known about the genes which make organisms useful and that it will be a long time before useful genetically engineered microbes are available. He advocates the wider use of proven microbial inoculants before attempting to use untested ones'.

(c) *Screening for competitive ability*

Competitive ability of a rhizobial strain against less effective indigenous strains is essential for successful establishment of the legume/*Rhizobium* symbiosis. Several methods based on serology, antibiotic resistant mutants, morphological or biochemical characteristics have been used to study competitive ability of strains (Somasegaran and Hoben, 1985). There are recently-developed methods, which can be used for this purpose based on the use of DNA probes, gene markers or macro-molecular patterns (Giller and Wilson, 1991). Each of these methods have advantages and disadvantages as summarized in Table 1.

The main disadvantage of all these methods is that they are time consuming. One has to analyse or identify bacteria in the nodules to be able to evaluate competitive ability of strains. There is one possibility around this problem using yield as an indication of competitive ability (Hardarson et al., 1981). If a legume plant is grown in nitrogen free medium and inoculated with a mixture of effective and ineffective rhizobia then the plant yield will depend on the nodulation success of the effective strain and its effective N_2 fixation. Strains which have been found to be similar for N_2 fixation effectiveness can be ranked for competitive ability only by measuring the field of the plants when the mixed strains are inoculated onto the legume. This method is illustrated by a fictitious example in Table 2. One can observe that strain E_2 is the most competitive against I followed by E_3, E_4, E_5 and E_1. The great advantage of this method is that nodule occupancy does not have to be analysed.

Breeding legumes for nitrogen fixation associated traits

(a) *FAO/IAEA Co-ordinated Research Programme on Enhancing Nitrogen Fixation in Common Bean*

The FAO/IAEA Coordinated Research Programme on 'The Use of Isotopes in Studies to Improve Yield and Nitrogen Fixation of Com-

Table 1. Methods for identification of micro-organisms

Advantages/ Disadvantages	Antibiotic resistant mutants	Serology	Molecular biology		
			PCR[a]	DNA probes	Genemarkers (GUS)
'Wildtype' strains	−	+	+	+	−
Simple and rapid, prep.	+	−	−	−	−
Simple and rapid, use	+	+	+	+	+
Specificity	+	−/+	+	+	+
Free or restricted use and transport	+	+	+	+	−
Low budget operation	+	+	+	−	+
Number of analyses day^{-1}	~500	~500	≦40	≦100?	≧500

[a] PCR = Polymerase chain reaction.

Table 2. Competitive screening of effective rhizobial strain

Strains	Relative yield produced by strains		Relative inhibition %	Competitive ranking
	Single strains	Mixture (Ex + I)		
E_1	100	20	100	5
E_2	95	95	0	1
E_3	90	70	28	2
E_4	95	50	60	3
E_5	85	30	85	4
I	20			

$$\text{Relative inhibition } (\%) = \frac{E - (I + E_{mix})}{E - I} \times 100$$

$$\text{for } E_4 = \frac{95 - 50}{95 - 20} \times 100 = 60\%$$

mon Bean (*Phaseolus vulgaris* L.) in Latin America' was initiated in 1986 and continued for five years. The objective of the programme was to increase common bean yield by increasing N derived from symbiotic fixation and by reducing their reliance upon soil or fertilizer N. The planning and research approach had been previously published (Hardarson and Lie, 1984). All participants performed similar types of experiments to quantify biological nitrogen fixation in approximately twenty cultivars or lines of common bean using the ^{15}N isotope dilution method. Most of the experiments yielded large variation between plant genotypes in terms of nitrogen fixation. In some of the experiments up to 65%N derived from atmosphere (% Ndfa) was observed, which is a much larger value than has previously been reported for common bean. Selected results from these experiments are summarized in Hardarson et al. (1993).

(b) *The programme at University of Wisconsin and CNPAF/EMBRAPA using the in-bred backcross line method*

The in-bred backcross line (IBL) method was developed to facilitate the transfer of genes controlling quantitatively inherited traits which often have low to moderate heritability. For those traits, the number of effective genes usually is not known and because of low heritability, selection of single plants is ineffective. The IBL method combines the important features of both backcrossing and the production of near-homozygous families (lines) by single seed descent (SSD). Following hybridization of the adapted cultivar (recurrent parent) and the source of high N_2 fixation (donor parent), the F_1 plant is backcrossed to the recurrent parent to produce a specified number of BC_1 individuals (~60 is a reasonable population size). Each BC_1 plant is backcrossed individually to the recurrent parent to produce ~60 unique BC_2 individuals. The population of ~60 BC_2 families is advanced by S.S.D. to the S_2 or S_3 generation, depending on the desired level of homozygosity, giving ~60 BC_2C_3 families. Enough seed is produced at this stage to allow replicated testing of the BC_2S_3 families.

Comparisons are made among segregating families, with the best families being compared to each other, the recurrent and donor parents or both for N_2 fixation and other traits. The magnitude of differences to be detected among families can be specified depending on the precision of the experiment and the amount of replication employed. The backcrosses to the recurrent parent accelerates recovery of the important essential features of the adopted cultivar which is being improved and assists in the evaluation of gene effects, since they are being transferred into a useful genetic background. The availability of near-homozygous families allows for replicated, multiple-trait evaluations on the same plant materials, i.e., the families (Bliss, 1993).

(c) *A multiple-trait breeding programme for improving the legume/*Rhizobium *symbiosis for N_2 fixation*

This programme was initiated at the University of Minnesota in 1974. It was the first concentrated effort in the United States to breed a forage legume for increased N_2 fixation (Barnes et al., 1981). The programme involved multidisciplinary research effort of scientists with expertise in plant breeding, plant physiology, soil microbiology, biochemistry, agronomy and plant pathology. The goals were mainly to: 1) develop an in-depth understanding of N_2 fixation in alfalfa, 2) breed alfalfa with an improved capability for N_2 fixation, and 3) develop strategies to apply N_2 fixation information and improved germplasm to agricultural production systems (Barnes et al., 1984).

The following four technologies from the programme of alfalfa improvement at the University of Minnesota have reached the developmental stage of field testing: 1) alfalfa germplasm with enhanced activities of nodule enzymes; 2) alfalfa germplasm unable to fix N_2; 3) alfalfa germplasm with improved N accumulation in herbage and roots, and 4) improved rhizobial strains for inoculum (Heichel et al., 1989).

(d) *Soybean genotypes selected for enhanced capacity to nodulate and fix nitrogen in the presence of nitrate*

This programme which has been conducted at the New South Wales Department of Agriculture has the main objective to select soybean genotypes which are able to nodulate and fix nitrogen in the presence of nitrate. The programme which screened almost 500 genotypes by the relative abundance of ureides in xylem exudate as an index of fixation was able to find four genotypes, all of Korean origin, showing high levels of symbiotic activity in the presence of nitrate (Herridge and Betts, 1988). The levels of nitrate tolerance in these lines may prove useful in breeding programmes which have the objective to increase nitrogen fixation in soybean in high N soils. A comparison of genes involved in the N tolerance of N_2 fixation in soybean to that of fababean mentioned earlier in this chapter should be a rewarding study.

(e) *The supernodulation soybean mutant project*

A supernodulating mutant nts382 was produced by EMS treatment of the parent cultivar Bragg at the Australian National University, Canberra (Carroll et al., 1985). The number of nodules on the tap root and on lateral roots were increased in the mutant line. In the presence of combined N nitrogenase activity was considerably higher in the mutant than in the parent cultivar. However the mutant plants were generally smaller than wild-type plants.

The supernodulating mutant has become a very useful research tool for the study of nodulation events and plant physiology in soybean. To become also agronomically important the mutation might have to be bred into other commercial cultivars with high yielding capacities.

(f) *The Brazilian programme on biological nitrogen fixation*

This programme conducted at several institutes in Brazil during 1940's to 1960's on the integrated study of soil, plant and bacterial involvement in N_2 fixation was among the strongest in Latin America. After the rhizobial inoculants had been selected, inoculation began to replace N fertilization in both experimental and farm crops. Since 1963 all experimental work on e.g. soybean has been based on inoculation rather than N fertilization. Current cultivars which can produce up to 4 tons ha^{-1} do not respond to N fertilizer except if inoculation has been unsuccessful (Döberiner, 1982).

The programme in Brazil is an example how nitrogen fixation can be enhanced parallel to improvement of other characteristics of grain legumes. One is likely to enhance nitrogen fixation if all selections, e.g. for yield, disease or stress resistance, are made without N fertilization but with rhizobial inoculation.

(g) *Breeding for rhizobial preference*

Preference for rhizobial strains has been found to be genetically controlled (Hardarson and Jones, 1979b). If one is to breed for rhizobial preference it is necessary to use a quick screening method (Hardarson et al., 1981). The screening can be done by inoculating several genotypes or breeding lines with a mixture of effective and ineffective strains under N deficient condition and measure the yield produced similar to the above method on screening for competitive ability. A fictitious example of this method is presented in Table 3. The advantage of the method is that no isolations of rhizobial strains have to be made.

Table 3. Screening of plant genotypes for preference for effective rhizobial strains

Plant genotype	Relative yield when inoculated with mixture of E and I strains	Ranking for preference for E strains
G_1	90	2
G_2	70	3
G_3	100	1
G_4	40	4

(h) *Soybean genotypes which reduce the competitiveness of the* Bradyrhizobium japonicum *strains serocluster 123*
The objective of this programme at the N_2 Fixation and Soybean Genetics Laboratory at Beltsville was the development of soybean genotypes which can eliminate nodulation by indigenous *Bradyrhizobium japonicum* of serocluster 123 without affecting the nodulation by other effective strains. Strains of serocluster 123 are very competitive in nodule formation but relatively ineffective in nitrogen fixation (Cregan et al., 1988). The programme was able to find several genotypes which restricted nodulation and altered competitiveness of strains USDA 123 (Keyser et al., 1988).

The above programmes as examples have concentrated efforts on the plant symbiont and clearly demonstrated that nitrogen fixation can be enhanced by breeding the plants for nitrogen fixation supportive traits. However like all breeding programmes they have to be given enough time and funds for improved varieties to be produced.

Conclusions

Despite fluctuations in the funding support for N_2 fixation research, it will likely continue to be of major interest in the agriculture of developing countries due to lack of N fertilizer and economical constraints, and in the developed countries due to increased awareness of groundwater pollution from chemical fertilizer and to attempt to create sustainable agriculture. However, if biological nitrogen fixation is to be utilized, it must be optimized. There are several methods available to the researcher when symbiotic nitrogen fixation must be enhanced, as shown in the present paper. No one approach is better than all others; rather work on symbiosis combining experience from various diciplines in interdisciplinary research programmes should be pursued.

References

Afza R, Hardarson G, Zapata F and Danso S K A 1987 Effects of delayed soil and foliar N fertilization on yield and N_2 fixation of soybean. Plant and Soil 97, 361–368.

Agboola A A and Fayemi A A A 1972 Fixation and excretion of nitrogen by tropical legumes. Agron. J. 64, 409–412.

Barnes D K, Heichel G H, Vance C P, Viands D R and Hardarson G 1981 Successes and problems encountered while breeding for enhanced N_2 fixation in alfalfa. *In* Genetic Engineering of Symbiotic Nitrogen Fixation and Conservation of Fixed Nitrogen. Eds. J M Lyons, R C Valentine, D A Phillips, D W Rains and R C Huffaker, pp. 233–248. Plenum Press, New York.

Barnes D K, Heichel G H, Vance C P and Ellis W R 1984 A multiple-trait breeding program for improving the symbiosis for N_2 fixation between *Medicago sativa* L and *Rhizobium meliloti*. Plant and Soil 82, 303–314.

Bergersen F J (Ed.) 1980 Methods for Evaluating Biological Nitrogen Fixation. Wiley, Chichester 701 p.

Beringer J E 1989 Biotechnology in relation to beneficial microorganisms. *In* Biotechnology Study Project Papers; Summaries of Commissioned Papers. World Bank, ISNAR, and ACIAR, Canberra, pp 30–31.

Bliss F A 1993 Breeding common bean for improved biological nitrogen fixation. Plant and Soil 152, 71–79.

Boote K J, Gallagher R N, Robertson W K, Hinson K and Hammond L C 1978 Effect of foliar fertilization on photosynthesis, leaf nutrition and yield of soybean. Agron. J. 70, 787–791.

Bowen G D and Danso S K A 1987 Nitrogen research for perennial crops. IAEA Bulletin 29, 5–8 p.

Brockwell J, Diatloff A, Roughly R J and Date R A 1982 Selection of rhizobia for inoculants. *In* Nitrogen Fixation in Legumes. Ed. J M Vincent pp 173–191. Academic Press, Sydney.

Carroll B J, McNeil D L and Gresshoff P M 1985 A supernodulation and nitrate-tolerant symbiotic (*nts*) soybean mutant. Plant Physiol. 78, 34–40.

Craswell E T 1990 Biological Nitrogen Fixation: Investments and Expectations from a Donor's Perspective Working Paper No. 30. ACIAR, Canberra 8 p.

Cregan P B, Keyser H H and Sadowsky 1988 Soybean genotypes which reduce the competitiveness of the *Bradyrhizobium japonicum* serotype strains comprising serocluster 123. *In* Nitrogen Fixation: Hundred Years After. Eds. H Bothe, F J de Bruijn and W E Newton. Gustav Fischer, Stuttgart 782 p.

Döbereiner J 1982 The Brazilian program in biological nitrogen fixation. *In* Biological Nitrogen Fixation Technology for Tropical Agriculture. Eds. P H Graham and S C Harris. pp 687–688. CIAT, Cali.

FAO 1984 Legume Inoculants and Their Use. Food and Agriculture Organization of the United Nations, Rome 63 p.

Danso S K A, Zapata F, Hardarson G and Fried M 1987 Nitrogen fixation in faba bean as affected by plant population density in sole or intercropped systems with barley. Soil Biol. Biochem. 19, 411–415.

Eaglesham A R J, Ayanaba A, Rao V R and Eskew D L 1981 Improving the nitrogen nutrition of maize by intercropping with cowpea. Soil Biol. Biochem 13, 169–171.

Gibson A H 1980 Methods for legumes in glasshouses and controlled environment cabinets. *In* Methods for Evaluating Biological Nitrogen Fixation. Ed. F J Bergersen pp 139–184. Wiley, Chichester.

Giller K E and Wilson K J 1991 Nitrogen Fixation in Tropical Cropping System CAB International 313 p.

Halliday J 1984 Integrated approach to nitrogen fixing trees germplasm development. Braz. J Agric. Res. 19, 91–117.

Hardarson G and Jones D G 1979a Effect of temperature on competition amongst strains of *Rhizobium trifolii* for nodulation of two white clover varieties. Ann. Appl. Biol. 92, 229-236.

Hardarson G and Jones D G 1979b The inheritance of preference for strains of *Rhizobium trifolii* by white clover (*Trifolium repens* L.). Ann. Appl. Biol. 92, 329–333.

Hardarson G and Lie T A 1984 Breeding Legumes for Enhanced Symbiotic Nitrogen Fixation. Martinus Nijhoff/ Dr. W. Junk Publishers, Dordrecht 166 p.

Hardarson G, Heichel G H, Vance C P and Barnes D K 1981 Evaluation of alfalfa and *Rhizobium meliloti* for compatibility in nodulation and nodule effectiveness. Crop Sci. 21, 562–567.

Hardarson G, Zapata F and Danso S K A 1984a Field evaluation of symbiotic nitrogen fixation by rhizobial strains using ^{15}N methodology. Plant and Soil 82, 369–375.

Hardarson G, Zapata F and Danso S K A 1984b Effect of plant genotype and nitrogen fertilizer on symbiotic nitrogen fixation by soybean cultivars. Plant and Soil 82, 397–406.

Hardarson G, Danso S K A and Zapata F 1987 Biological nitrogen fixation in field crops. *In* Handbook of Plant Science in Agriculture Ed. B R Christie pp 165–192. CRC Press Inc., Boca Raton, FL.

Hardarson G, Danso S K A and Zapata F 1988 Dinitrogen fixation measurement in alfalfa-ryegrass swards using nitrogen-15 and influence of the reference crop. Crop Sci. 28, 101–105.

Hardarson G, Golbs M and Danso S K A 1989 Nitrogen fixation in soybean (*Glycine max* L. Merrill) as affected by nodulation patterns. Soil Biol. Biochem. 21, 783–787.

Hardarson G, Danso S K A, Zapata F and Reichardt K 1991 Measurements of nitrogen fixation in fababean at different N fertilizer rates using the ^{15}N isotope dilution and 'A-value' methods. Plant and Soil 131, 161–168.

Hardarson G, Bliss F A, Cigales-Rivera M R, Henson R A, Kipe-Nolt J A, Longeri L, Manrique A, Peña-Cabriales J J, Pereira P, Sanabria C A and Tsai S M 1993 Genotypic variation in biological nitrogen fixation by common bean. Plant and Soil 152, 59–70.

Heichel G H, Barnes D K, Vance C P and Sheaffer C C 1989 Dinitrogen fixation technology for alfalfa improvement. J. Prod. Agric. 2, 24–32.

Herridge D F and Betts J H 1988 Field evaluation of soybean genotypes selected for enhanced capacity to nodulate and fix nitrogen in the presence of nitrate. Plant and Soil 110, 129–135.

IITA 1985 Grain Legume Improvement Program: Research Highlights 1981–1984. IITA, Nigeria. 83 p.

Keyser H H, Cregan P B and Sadowsky M J 1988 Soybean genotypes which restrict nodulation of strains of *Bradyrhizobium japonicum* serocluster 123. *In* Nitrogen Fixation: Hundred Years After. Eds. H Bothe, F J de Bruijn and W E Newton. Gustav Fischer, Stuttgart 784 p.

Kuykendall L D and Weber D F 1978 Genetically marked *Rhizobium* identifiable as inoculum strain in nodules of soybean plants grown in field populated with *Rhizobium japonicum*. Appl. Environ. Microbiol. 36, 915–919.

McDermott T R and Graham P H 1989 *Bradyrhizobium japonicum* inoculant mobility, nodule occupancy, and acetylene reduction in the soybean root system. Appl. Environ. Microb. 55, 2493–2498.

Peoples M B, Faizah A W, Rerkasem B and Herridge D F 1989 Methods for Evaluating Nitrogen Fixation by Nodulated Legumes in the Field. ACIAR, Canberra 76 p.

Poole W D, Randall G W and Ham G E 1983 Foliar fertilization of soybean. I. Effect of fertilizer source, rates and frequency of application. Agron. J. 75, 195–200.

Ramos M L G, Ribeiro W O Jr and Kipe-Nolt J A 1993 Effect of fungicides on survival of Rhizobium on seeds and the nodulation of bean (*Phaseolus vulgaris* L) Plant and Soil 152, 145–150.

Rosas J C and Bliss F A 1986 Host plant traits associated with estimates of nodulation and nitrogen fixation in common bean. Hort Science 21, 287–289.

Roughley R J 1980 Environmental and cultural aspects of the management of legumes and *Rhizobium*. *In* Advances in Legume Sciences Eds. R J Summerfield and A H Bunting. pp. 97-103. Royal Botanic Gardens, Kew.

Roughley R J, Sprent J I and Day J M 1983 Nitrogen fixation. *In* The Faba Bean (*Vicia faba* L.) Ed. D D Habblethwaite. pp 230–260. Butterworths, London.

Senaratne R, Amornpimol C and Hardarson G 1987 Effect of combined nitrogen on nitrogen fixation of soybean (*Glycine max* L. Merrill) as affected by cultivar and rhizobial strain. Plant and Soil 103, 45–50.

Silvester-Bradley R, Ayarza M A, Mendez J E and Moriones R 1983 Use of undisturbed soil cores for evaluation of *Rhizobium* strains and methods for inoculation of tropical forage legumes in a Colombian Oxisol. Plant and Soil 74, 237–247.

Somasegaran P and Hoben H J 1985 Methods in Legume-Rhizobium Technology. NifTAL, MIRCEN, Hawaii 367 p.

Tauer L W 1989 Economic impact of future biological nitrogen fixation technologies on United States agriculture. Plant and Soil, 119, 261–270.

Viands D R, Barnes D K and Heichel G H 1981 Nitrogen Fixation in Alfalfa: Responses to Bidirectional Selection for Associated Characteristics. Technical Bulletin No. 1643. USDA-ARS 18 p.

Vincent J M 1970 A Manual for the Practical Study of Root-Nodule Bacteria. Blackwell Scientific Publications, Oxford 164 p.

Wolyn D J, Attewell J, Ludden P W and Bliss F A 1989 Indirect measure of N_2 fixation in common bean (*Phaseolus vulgaris* L.) under field conditions. Role of lateral root nodules. Plant and Soil 113, 181–187.

Zapata F, Danso S K A, Hardarson G and Fried M 1987 Time course of nitrogen fixation in field-grown soybean using nitrogen-15 methodology. Agron. J. 79, 172–176.

Plant and Soil **152**: 19–23, 1993.

PLSO BNF02

Methods for measuring biological nitrogen fixation in grain legumes

G. HARDARSON and S.K.A. DANSO
FAO/IAEA Programme, Soil Science Unit, Agency's Laboratories, A-2444 Seibersdorf, Austria and Joint FAO/IAEA Division, P.O. Box 100, A-1400 Vienna, Austria

Key words: acetylene reduction, difference method, legumes, ^{15}N isotope method, nitrogen, nitrogen fixation, *Rhizobium*

Abstract

To assure proper management and fully realize the benefits of the legume-*Rhizobium* symbiosis it is necessary to be able to quantify the amount of nitrogen fixed. Having measured the effectiveness of atmospheric N_2 fixation the macro- or micro-symbionts as well as agronomic factors can be manipulated with the objective to maximize biological nitrogen fixation. A suitable method to quantify nitrogen fixation is therefore necessary in any programme aiming at increasing N_2 fixation, like the one being reported in this volume. There are several methods available to quantify nitrogen fixation and most of the commonly used ones are described in the present paper listing their advantages and disadvantages.

Introduction

An important characteristic of legumes is their ability to fix atmospheric nitrogen in symbiosis with nodule-forming *Rhizobium* bacteria. Legumes undergoing effective biological nitrogen fixation can therefore be grown with less applied nitrogen. Having this characteristic, they are particularly important in countries where the cost of nitrogen fertilizer is high and/or availability is limited.

To assure proper management and fully realize the benefits of this plant-microbial association, it is necessary to estimate the amount of nitrogen fixed under different field conditions. It is only after this is known that various factors can be manipulated to increase the amount and proportion of N the legume plant derives from the atmosphere. A suitable method for accurately measuring the amount of nitrogen fixed is an important requirement in any programme aimed at maximizing symbiotic N_2 fixation.

There are several methods available, all having some advantages and disadvantages. The most suitable method depends largely on the objective of the experiment. Methods for estimating N_2 fixation have been described and discussed in several publications (Bergersen, 1980; Hardarson et al., 1987; Hardarson, 1990; Peoples et al., 1989; Vincent, 1970). Some of these will be described only briefly here and reference will be made to other publications where more details are presented.

Methods for measuring nitrogen fixation

Dry matter yield method (DM)

If nitrogen is the primary limiting factor for growth, dry matter yield of the legume plant should be positively correlated with the amount of N from fixation. This method is often used to screen large numbers of rhizobial strains and host plant lines. It has the advantage of being simple and cheap to carry out; and non-destructive if only fresh weight is being recorded and the plants can be replanted for seed production.

Total N difference method

This method is related to the previous one with the additional quantification of % N in the fixing plants and non-fixing reference crop plants. Although the need for Kjeldahl N analyses (Eastin, 1978) of the samples makes this method more time consuming, it provides more information on the amount of nitrogen fixed. Recently infrared reflectance methods calibrated against Kjeldahl N have been used to determine plant N, resulting in considerable time savings and similar accuracy. The main disadvantage of this method is that legume and reference crops must absorb similar amounts of nitrogen from the soil. When this is not the case erroneous estimates of N_2 fixation may result. Further information on this method can be obtained from Peoples et al. (1989).

Nodule observations

Nodule number and weight have often been found to be positively correlated to the amount of N_2 fixed. Such measurements (or visual scoring) can be useful when large numbers of rhizobial strains and plant germplasm are to be evaluated. Nodule observations when collected in conjunction with other measurements are especially useful for the interpretation of data.

Acetylene reduction assay (ARA)

In this method nodulated roots are incubated in acetylene and ethylene produced after a specific time as measured by gas chromatograph. Although simple to use, it provides only point measurements of nitrogenase activity as measured by gas chromatograph. It is not advisable to calculate N_2 fixation from these measurements since the calibration factors are variable for different fixing systems and also variable with time. Further errors may arise due to acetylene-induced decline in nitrogenase activity during assay (Witty and Minchin, 1988). Additional information on this method may be obtained from Bergersen (1980).

Xylem-solute technique

This method is based on measurements and comparisons of N-containing compounds, which are transported from the roots to the shoots in the xylem sap. These compounds are either assimilation products of N_2 fixing nodules or mineral N taken up from soil. When there are differences in xylem N-solute composition between fixing and non-fixing crops of the same species then analysis of xylem sap may be used to assess nitrogen fixation indirectly. Many legumes of tropical origin (e.g., soybean, common bean and cowpea) transport most of their fixed N in the form of ureides, allantoin and allantoic acid. The xylem-solute technique is applicable mainly for these legumes. Other legumes (e.g. groundnuts, lupins, faba bean, alfalfa and clovers) export fixed N_2 mainly as amides, asparagine and glutamine. The response of these legumes to changes in N-source is less well defined than the ureide-producing species. Therefore the method is used predominantly for ureide-producing plants. Further information on this technique can be obtained from Peoples et al. (1989).

^{15}N methodologies

a) Use of $^{15}N_2$ gas

The earliest use of $^{15}N_2$ to study nitrogen fixation was by Burris and Miller (1941). This method uses ^{15}N enriched gas in which nitrogen fixing systems are incubated. It provides direct evidence for N_2 fixation since the ^{15}N concentration in plants exposed to $^{15}N_2$ is greater than the 0.3663% natural abundance if fixation has occurred. The extent to which ^{15}N is detected in the plant provides an estimate of the proportion of the plant's N derived from the atmosphere during the incubation time. The environment within the chamber is, however, different from that in a field situation, and the plants cannot be confined in these chambers for long periods. Results obtained from such studies tend to be instantaneous and subject to errors associated with extrapolating data from short-term studies to an entire growing season which involves diurnal, daily and seasonal variations (Knowles, 1980).

b) Use of ^{15}N enriched fertilizers or substrates – Isotope dilution method (ID)

This method involves the growth of N_2 fixing and non-fixing reference plants in soil fertilized with ^{15}N enriched inorganic or organic fertilizers. It is based on differential dilution in the plant of ^{15}N-labelled fertilizer by soil and fixed nitrogen (Fried and Middelboe, 1977; McAuliffe et al., 1958). The use of ^{15}N labelled fertilizer provides an integrated measurement of the amount of fixed N accumulated by a crop over the growing season. There are two main variations of this isotope method i.e. the isotope dilution method and the A-value method.

In the isotope dilution method both fixing and non-fixing reference plants are grown in soil to which the same amount of fertilizer having the same ^{15}N enrichment has been applied. In the absence of any supply of N other than soil and ^{15}N-labelled fertilizer, both plants will contain the same ratio of $^{15}N/^{14}N$, since they are taking N of similar $^{15}N/^{14}N$ composition, but not necessarily the same total quantity of N. However, in the presence of N_2, the fixing plant further lowers the ratio of $^{15}N/^{14}N$ due to incorporation of N from unlabelled air, while this does not occur in the reference plant. Calculation of % N derived from atmosphere (% Ndfa) can be made by the following equation:

$$\%\ \text{Ndfa} = \left(1 - \frac{\%\ \text{Ndff}_F}{\%\ \text{Ndff}_{NF}}\right) \times 100 \qquad (1)$$

where $\%\ \text{Ndff}_F$ and $\%\ \text{Ndff}_{NF}$ are % N derived from fertilizer by fixing and non-fixing plants, respectively.

The equation (1) is derived from the following ones as shown by Fried and Middelboe (1977) and Hardarson (1990):

$$\%\ \text{Ndff}_{NF} + \%\ \text{Ndfs}_{NF} = 100 \qquad (2)$$

$$\%\ \text{Ndff}_F + \%\ \text{Ndfs}_F + \%\ \text{Ndfa} = 100 \qquad (3)$$

and

$$\frac{\%\ \text{Ndff}_{NF}}{\%\ \text{Ndfs}_{NF}} = \frac{\%\ \text{Ndff}_F}{\%\ \text{Ndfs}_F} \qquad (4)$$

where $\%\ \text{Ndfs}_F$ and $\%\ \text{Ndfs}_{NF}$ are the percentage N derived from soil in the fixing and non-fixing crops, respectively. Equation (4) represents the only assumption made in the ^{15}N isotope dilution method i.e. that both non-fixing and fixing crops take up N from soil and fertilizer in the same ratio. For this to be true the fixing and non-fixing crops must match, i.e., (a) either the fertilizer distribution is even with depth or the legume and reference crops have similar root systems and spatially similar nutrient uptake profiles, and (b) the enrichment of substrate remains constant with time or the legume and reference crops have similar N uptake patterns (Witty, 1984).

The ^{15}N isotope dilution method was used extensively in the programme presented in this volume for both greenhouse and field studies. In each case equal amounts and enrichments of ^{15}N-labelled fertilizer were applied to the fixing and non-fixing reference crops usually at the time of planting. Percentage and amount of N derived from the atmosphere were evaluated at physiological maturity before leaf senescence of the legume crop.

c) A-value method (AV)

In the A-value method different rates of N fertilizer are applied to the fixing and non-fixing plants. It may be necessary to apply a higher rate to the non-fixing reference crop to support its proper growth, especially in low fertility soils. However, increased levels of inorganic N can depress nitrogen fixation. For this reason it is practical to give a reasonable dose of ^{15}N labelled fertilizer (40–80 kg N ha^{-1}) to the reference crop, while the fixing crop receives a low quantity (5–20 kg N ha^{-1}) (Fried and Broeshart, 1975). % Ndfa may be calculated using the following equation:

$$\%\ \text{Ndfa} = 100\left(1 - \frac{\%\ \text{Ndff}_F}{n\%\ \text{Ndff}_{NF}}\right) + \%\ \text{Ndff}_F\left(\frac{1}{n} - 1\right) \qquad (5)$$

where n is the relative amount of fertilizer

Table 1. Comparison of methods to quantify symbiotic nitrogen fixation

	DM	Total N	Nods	ARA	Xylem	$^{15}N_2$	ID	AV	NA
Direct	–	–	–	–	–	+	–	–	–
No ref. plant needed	–	–	+	+	+	+	–	–	–
Simple, rapid and cheap	+ +	+	+	+ +	+	–	–	–	–
Non-destructive	–(+)	–	+	+	(+)	–	–	–	–
Time integrated measure	+	+	+	–	–	–	+	+	+
% Ndfa measure	–	+	–	–	–	–	+ +	+ +	+
Measure of kg/ha fixed	–	+	–	–	–	–	+ +	+ +	+
Small field variability	+	+	–	–	?	–	+	+	?

DM: Dry matter yield method
Total N: Total N difference method
Nods: Nodule observation
ARA: Acetylene reduction assay
Xylem: Xylem-solute technique
$^{15}N_2$: Use of $^{15}N_2$ gas
ID: ^{15}N isotope dilution
AV: A-value method
NA: ^{15}N natural abundance method

applied, i.e., n = amount of N fertilizer applied to the F crop divided by the amount applied to the NF crop (Hardarson et al., 1991). Further information on the isotope dilution and A-value methods can be obtained from Hardarson (1990).

d) Natural abundance method (NA)
As a result of isotope discrimination effects that occur during soil formation, most soils have slightly higher ^{15}N abundance than the atmosphere. Because of this difference in ^{15}N abundance between soil and atmospheric N_2, nitrogen-fixing plants have been found to have a lower ^{15}N enrichment than non-fixing ones; therefore this has been used to measure biological nitrogen fixation (Amarger et al., 1979; Kohl et al., 1980). Additional information on this method may be obtained from Bergersen (1980) and Peoples et al. (1989).

Selection of a method for quantifying symbiotic nitrogen fixation

A comparison of the characteristics of the various methods for measuring biological nitrogen fixation is shown in Table 1. Only by the use of the $^{15}N_2$-gas method can nitrogen fixation be measured directly. Some methods will require non-fixing reference crops, which may be considered as a disadvantage.

The dry matter yield, total N, ARA and xylem-solute methods are simple, rapid and relatively cheap. Therefore they may be used in breeding programmes were many analyses must be performed on segregating plant populations and precision is of secondary importance.

When selecting plants for N_2 fixation related traits, it may be important to evaluate plants non-destructively for later production of seed from these same plants. For those plants that can be transplanted, ARA, xylem-solute and in this case fresh instead of dry matter yield methods may be used for this purpose.

Time integrated measurement of N_2 fixation and quantification of % Ndfa and total Ndfa may be important only at the beginning of a breeding programme, when gene donors and recipients must be identified accurately, and at the final stage of the programme when the resulting improvements must be measured to determine progress and suitability for release to the grower. Total N, ID, AV and NA methods will be the most suitable for these purposes, when only few and relatively accurate measurements have to be made. Uniform field sites will be important in these studies.

References

Amarger N, Mariotti A, Mariotti F, Durr J C, Bourguignon C and Lagacherie B 1979 Estimate of symbiotic fixed nitrogen in field grown soybean using variations in ^{15}N natural abundance. Plant and Soil 52, 269–280.

Bergersen F J 1980 Methods for Evaluating Biological Nitrogen Fixation. Wiley, Chichester. 701 p.

Burris R H and Miller C E 1941 Application of ^{15}N to the study of biological nitrogen fixation. Science 93, 114–115.

Eastin E F 1978 Total nitrogen determination for plant material. Anal. Biochem. 85, 591–594.

Fried M and Broeshart H 1975 An independent measurement of the amount of nitrogen fixed by legume crops. Plant and Soil 43, 707–711.

Fried M and Middelboe V 1977 Measurement of amount of nitrogen fixed by a legume crop. Plant and Soil 47, 713–715.

Hardarson G, Danso S K A and Zapata F 1987 Biological nitrogen fixation in field crops. *In* Handbook of Plant Science in Agriculture. Ed. B R Christie. pp 165–192. CRC Press Inc., Boca Raton, FL.

Hardarson G Ed. 1990 Use of Nuclear Techniques in Studies of Soil-Plant Relationships. Training Course Series No. 2, IAEA, Vienna. 223 p.

Hardarson G, Danso S K A, Zapata F and Reichardt 1991 Measurements of nitrogen fixation in fababean at different N fertilizer rates using the ^{15}N isotope dilution and 'A-value' methods. Plant and Soil 131, 161–168.

Knowles R 1980 The measurement of nitrogen fixation. *In* Current Perspectives in Nitrogen Fixation. Proceedings of the Fourth International Symposium on Nitrogen Fixation. Eds A H Gibson and W E Newton. pp 327–333. Elsevier, North-Holland Biomedical Press, Amsterdam.

Kohl D H, Shearer G and Harper J E 1980 Estimates of N_2 fixed based on differences in natural abundance of ^{15}N in nodulating and non-nodulating isolines of soybeans. Plant Physiol. 66, 61–65.

McAuliffe C, Chamblee D S, Uribe-Arango H and Woodhouse W W 1958 Influence of inorganic nitrogen on nitrogen fixation by legumes as revealed by ^{15}N. Agron. J. 50, 334–337.

Peoples M B, Faizah A W, Rerkasem B and Herridge D F 1989 Methods for Evaluating Nitrogen Fixation by Nodulated Legumes in the Field. Australian Centre for International Agricultural Research, Canberra, 1989. 76 p.

Vincent J M 1970 A Manual for the Practical Study of Root-Nodule Bacteria. IBP Handbook No 15. Blackwell Scientific Publications, Oxford. 164 p.

Witty J F 1984 The validity of some assumptions inherent in the application of the acetylene reduction assay and isotope dilution method. Biological N_2 Fixation Newsletter 12, 1–3.

Witty J F and Minchin F R 1988 Measurment of nitrogen fixation by the acetylene reduction assay; myths and mysteries. *In* 'Nitrogen Fixation by Legumes in Mediterranean Agriculture'. Eds. D P Beck and L A Materon. pp 331–344. Martinus Nijhoff Publishers, Dordrecht.

Plant and Soil **152**: 25–52, 1993.

PLSO BNF03

Misconceptions and practical problems in the use of ^{15}N soil enrichment techniques for estimating N_2 fixation

S.K.A. DANSO[1], G. HARDARSON[2] and F. ZAPATA[2]
[1]*Joint FAO/IAEA Division, P.O. Box 100, A-1400 Vienna, Austria*; [2]*Soil Science Unit, Joint FAO/IAEA Programme, Agency's Laboratories, A-2444 Seibersdorf, Austria*

Key words: A-value method, legumes, ^{15}N isotope dilution, nitrogen fixation, *Rhizobium*

Abstract

The ^{15}N methods are potentially accurate for measuring N_2 fixation in plants. The only problem with those methods is, how to ensure that the $^{15}N/^{14}N$ ratio in the plant accurately reflects the integrated $^{15}N/^{14}N$ ratio (R) in soil which is variable in time and with soil depth. However, the consequences of using an inappropriate reference plant vary with the level of N_2 fixation and the conditions under which the study was made. For example, the errors introduced into the values of N_2 fixation are higher at low levels of fixation, and decrease with increasing rates of fixation. At very high N_2 fixation rates, the errors are often insignificant. Also, the magnitude of error is proportional to the rate of decline of the $^{15}N/^{14}N$ ratio with time. Since N_2 fixation in most plants would be expected to below 60%, the question of how to select a good reference plant is still pertinent. In this paper, we have discussed some of the criteria to adopt in selecting reference plants, e.g. how to ensure that the reference plant is not fixing N_2, is absorbing most of its N from the same zone as the fixing plant, and in the same pattern with time, etc. In addition, we have discussed ^{15}N labelling materials and methods that are likely to minimize any errors even when the fixing and reference plants don't match well in certain important criteria. The use of slow release ^{15}N fertilizer or ^{15}N labelled plant materials results in slow changes in the $^{15}N/^{14}N$ ratio of soil, and is strongly recommended. Where ^{15}N inorganic fertilizers are used, the application of the fertilizer in small splits at various intervals is recommended over a one-time application. The problem with the reference crop, which has sometimes discouraged potential users of the ^{15}N methods, is surmountable, as discussed in this paper.

Introduction

The measurement of atmospheric nitrogen (N_2) fixation in plants is an important step in efforts to increase the contribution of atmospheric N_2 to plant nutrition and soil fertility. Although there are several methodologies for measuring N_2 fixation in plants, none of them measures N_2 fixed with absolute accuracy. The advantages and limitations of each method have been discussed in several reviews (Bremner, 1977; Burris, 1974; Danso, 1985; Hardy et al., 1973; and Knowles, 1981). However, of the methods currently available, the indications are that the ^{15}N soil-enrichment techniques are the most reliable for quantifying N_2 fixed in field-grown crops (Duhoux and Dommergues, 1985; Duque et al., 1985; Hardarson et al., 1984b; Legg and Sloger, 1975; Rennie, 1982; 1984; Ruschel et al., 1979; 1982; Sanginga et al., 1985; Talbott et al., 1982; Vose et al., 1982; West and Wedin, 1985). These ^{15}N techniques are particularly useful because they can at a single harvest, measure the integrated amounts of N_2 assimilated by both greenhouse and field-grown plants in addition to measuring the N contributed from soil or fertilizer sources (Danso 1985; Fried et al., 1983; Vose and Victoria, 1986). They are thus invaluable in programmes that aim primarily at increasing the N_2, rather than soil or fertilizer N input for high yields. Consequently, the literature reveals an increasing use of the ^{15}N isotope techniques for measuring N_2 fixed in various crops and cropping systems (Chalk, 1985; Ledgard et al., 1985; Weaver, 1986).

The two broad approaches of the ^{15}N techniques for measuring N_2 fixed by isotope dilution depend on: (1) the usually slightly higher $^{15}N/^{14}N$ ratio of soil N (R) than atmospheric N_2, known as the natural ^{15}N abundance technique (Amarger et al., 1979; Kohl and Shearer, 1980), or (2) the magnification of these differences in R by adding ^{15}N-enriched N fertilizers to soil (Fried and Broeshart, 1975; Fried and Middelboe, 1977; McAuliffe, 1958; Vallis et al., 1967). Both approaches are, in essence, based on the same principle, i.e., the R absorbed by the N_2-fixing plant is decreased by the lower ^{15}N enrichment of atmospheric N_2. Thus the more N_2 that is fixed, the greater is the dilution of R. In a few cases, ^{15}N-depleted rather than ^{15}N-enriched fertilizers have been added to create a difference in R between soil N and atmospheric N_2 (St. Claire et al., 1988). Because the ^{15}N-enriched fertilizer approach requires less sophisticated analytical capability, and is subject to fewer errors than the natural ^{15}N abundance approach (Ledgard et al., 1985b), it is the most commonly used in N_2 fixation studies. The ^{15}N natural abundance method on the other hand is becoming increasingly popular especially in measuring N_2 fixation in undisturbed ecosystems with nitrogen fixing trees. The ^{15}N-depleted fertilizer approach is used infrequently, because the amount of fertilizer needed to create a significant ^{15}N difference between soil N and atmospheric N_2 may often be inhibitory to N_2 fixation.

The increased adoption of ^{15}N techniques in N_2-fixation research has on the other hand revealed some practical problems as well as some improper use, thereby prompting some criticisms and close examination of the methodology (Chalk, 1985). However, some of these limitations seem to have been amplified, and also, as suggested by Vose and Victoria (1986), a major problem with the use of ^{15}N techniques has been a general lack of understanding of some of the underlying concepts. It is therefore likely that with better understanding of the concepts, and adequate precautions to reduce some of the avoidable errors, many of the reported deficiencies could be substantially reduced. This paper addresses some of the reported problems and controversies of the ^{15}N soil enrichment techniques, and discusses attempts or possibilities to improve their efficacy.

Reference crops

Basis for requirement

The ^{15}N enrichment in a N_2-fixing plant is the end result of the assimilation of native N from soil, N_2 from the atmosphere, and N from the applied ^{15}N fertilizer. Because native soil N and atmospheric N_2 have almost similar ^{15}N enrichments, it is not easy to directly distinguish native soil N from fixed N_2, and to therefore estimate R in the fixing plant. Even where ^{15}N enriched fertilizer is added to soil, R cannot be assessed by instantaneous estimations of $^{15}N/^{14}N$ ratios in soil extracts, because (1) R in a ^{15}N fertilized soil normally changes with time (Fried et al., 1983; Henzell et al., 1968; Vallis et al., 1967; Witty, 1983) and (2) unequal quantities of this changing R are absorbed at different times during plant growth, making even an arithmetic mean of several assessments invalid. A reference non-fixing crop, which derives all its N from the soil however provides an integrated value for the ratio in which soil N and ^{15}N-labelled fertilizer were available during the measurement period. This does not, however, imply that the R estimated by any non-N_2-fixing crop is valid for any N_2-fixing crop. Certain criteria must be met (Fried et al., 1983). The selection of an appropriate reference crop for a particular fixing crop is therefore often necessary, and fulfilling this requirement frequently appears to be the greatest problem encountered with the ^{15}N methodology.

For the R estimated by the reference non-fixing crop (nfc) to be valid for the fixing crop (fc), it is necessary that both crops assimilate the ^{15}N fertilizer and native soil N in the same proportion throughout growth. This requirement can be expressed mathematically as follows:

$$\frac{\%\ \text{Ndff (nfc)}}{\%\ \text{Ndfs (nfc)}} = \frac{\%\ \text{Ndff (fc)}}{\%\ \text{Ndfs (fc)}} = \text{R} \qquad (1)$$

Where % Ndff and % Ndfs are the proportions of N derived from the fertilizer and soil, respectively. Accordingly, differences in total N absorbed from the soil by the fixing and reference crop are unimportant, so far as the integrated ratio in which these two N sources is absorbed is similar for both crops. For equation (1) to hold, (i) the reference crop should not fix N_2, (ii) both the fixing and non-fixing crops must absorb N from the same nutrient pool, and (iii) in similar ratios of fertilizer N and soil N during each phase of growth. (Danso, 1985; Fried et al., 1983; Witty, 1983). The extent of the dilution of R (assessed by the reference crop) in the fixing crop is then a valid measure of the proportion of N_2 fixed (Fried and Broeshart, 1975; Fried and Middelboe, 1977; McAuliffe et al., 1958; Vallis et al., 1967). Non-adherence to the above essential requirements during the choice of the reference crop has undoubtedly contributed enormously to serious errors in estimations of N_2 fixed. This is illustrated by reports of many significantly different estimates of N_2 fixed in a treatment in which different reference crops were used (Ledgard et al., 1985a; Rennie, 1982; Rennie et al., 1982; Wagner and Zapata, 1982; Witty and Ritz, 1984), indicating that the reference crops assessed different R values. Also, the reported negative values of N_2 fixed (Giller and Witty, 1987; Ledgard et al., 1985e; Patterson and LaRue, 1983; Reichardt et al., 1987; Vasilas and Ham, 1984; Wagner and Zapata, 1982 and Witty and Ritz, 1984) are unrealistic, and only indicate the magnitude of errors associated with these estimations. An example is the study of Witty and Ritz (1984), in which using maize as a reference crop estimated negative N_2 fixed in soybean, but changed to positive values when practices that resulted in less reference crop-induced errors were adopted, or when a suitable reference crop such as non-nodulating soybean was used. Describing the need for a reference crop as a limitation of the methodology, however, gives the false impression that it is impossible to meet the necessary requirements, which have been achieved in many studies (e.g., Fried et al., 1983; Rennie, 1982; Rennie and Dubetz, 1986; Wagner and Zapata, 1982; Witty and Ritz, 1984). What is needed whenever it is necessary to quantify N_2 fixed is, greater attention in choosing the reference crop.

Establishing the need for a reference crop

There seems to be an impression that the ^{15}N soil enrichment techniques always demand the use of rigorously selected reference crops. We shall examine the methodology in terms of (1) when there is absolute need for a reference crop and (2) when the errors introduced by a mismatch between the reference and fixing crop would be crucial.

As suggested by Phillips et al. (1986), the greatest utility of the ^{15}N isotope dilution assay lies in its usefulness to discern relative differences in N_2 fixation capabilities. The determination of the actual amount of N_2 fixed is necessary only in some instances, such as in N-balance studies in soil or the soil-plant system, or for comparing N_2 fixed in different seasons, years or environments. These considerations are examined below in more detail.

Non-obligatory requirement for reference crops

According to Danso (1986), Danso et al. (1986), Phillips et al. (1986), the need for the precise quantification of N_2 fixation may not be compelling in many agronomic trials, e.g., in comparing some treatment effects, or ranking varieties or *Rhizobium* strains for N_2 fixing effectiveness. This point may be clarified by examining the equation for estimating the fraction of N derived from atmospheric N_2 fixation (fNdfa):

$$\text{fNdfa} = 1 - \frac{\%\ ^{15}\text{N a.e. (fc)}}{\%\ ^{15}\text{N a.e. (nfc)}} \qquad (2)$$

Where fc and nfc are the fixing and non-fixing crops, respectively, and a.e. is atom excess. Thus, in using a given reference crop to compare N_2 fixed by two or more crops, the % ^{15}N a.e. in the reference crop is common, and therefore a constant value. The differences in N_2 fixation are then only dependent upon the relative ^{15}N enrichments in the various fixing crops. Therefore, with an identical N rate of a given ^{15}N-labelled fertilizer applied to various fixing crops or treatments (assumed or shown to absorb N from the same zone and in a similar pattern), the % ^{15}N a.e. differences in the plants are satisfactory indicators of the relative differences in N_2 fixing abilities (Hardarson et al., 1984). Based on the ^{15}N enrichment therefore, it should be possible to rank which treatments fixed more N_2 than others; if this is the only important question, then it is superfluous to use a reference crop, which could above all, introduce unnecessary errors (especially if a wrong one is selected) and also increase the size and cost of the study. An example to illustrate the usefulness of ^{15}N enrichment data in relative ranking is presented in Table 1 in which a non-nodulating soybean isoline was used to estimate % Ndfa in N_2-fixing soybean varieties inoculated with two *Rhizobium* strains (Danso et al., 1987). Using a reference crop indicated that Amsoy variety nodulated by a Nitragin strain obtained the highest % Ndfa, and Chippewa, nodulated by strain 'D' the least. Without considering the reference crop, the same ranking as that obtained by the quantitative estimation (i.e., using the reference crop) is made by simply using the ^{15}N enrichment data (Table 1), on the basis that the lower the % ^{15}N a.e. in the plant, the higher the N_2 fixed. These ^{15}N differences among treatments could even on several occasions give approximate and yet reliable indications of the magnitude to which each treatment influenced N_2 fixation. For example, the large difference in % Ndfa between, for example, Amsoy (65%) and Chippewa (10%) (Table 1) is also reflected in a correspondingly big difference in % ^{15}N a.e. (0.0456 for Amsoy and 0.1112 for Chippewa inoculated with strain 'D'). In another example, Ruschel et al. (1982) compared the amounts of N_2 fixed by *Phaseolus* varieties harvested at either 60 or 90 days after planting (DAP), when the reference crop was harvested only once, with the 60-day *Phaseolus* harvest. There was therefore no valid reference crop for the 90-day varieties, and thus neither the estimated N_2 fixed in the later-maturing varieties nor their comparison with the 60-day varieties is valid. These comparisons would however have been valid and more meaningful if they had simply been restricted to an examination of the relative ^{15}N enrichments or % Ndff within only the legumes harvested at 90 DAP (even though they had no reference crop). Furthermore, Hardarson et al. (1984b) and Labandera et al. (1988), indicated that direct ^{15}N isotopic parameters, such as % ^{15}N a.e. or % Ndff in fixing crops are more sensitive and perhaps result in more accurate relative comparisons than other isotope-derived estimates (e.g., those involving reference crops). Reliance on ^{15}N enrichments in fixing crops could therefore be very useful in discerning small differences in N_2 fixation in wide-scale programmes aimed at selecting or breeding plant genotypes for higher N_2-fixing abilities, and also, in *Rhizobium* strain evaluation trials. The accurate ranking of N_2 fixing abilities based on relative ^{15}N enrichments however requires the selection of fairly homogeneous soil sites (Hardarson et al., 1984; Reichardt et al., 1987), and the use of the same amount and enrichment of a given ^{15}N fertilizer on all crops under examination.

In some extremely difficult situations such as with sugarcane, for which the selection of an appropriate reference crop, of similar growth duration has not been easy (Vose and Victoria, 1986), a modified relative ^{15}N enrichment ap-

Table 1. Relative comparison of treatment effects on N_2 fixation, using either % ^{15}N enrichment in legumes or % Ndfa values

Soybean variety	*Rhizobium* strain	% ^{15}N atom excess	% Ndfa
Amsoy	Nitragin	0.0456 (1)*	65 (1)*
Williams	'D'	0.0552 (2)	56 (2)
Williams	Nitragin	0.0572 (3)	54 (3)
Chippewa	Nitragin	0.0885 (4)	28 (4)
Chippewa	'D'	0.1112 (5)	10 (5)

* Figures in parenthesis represent relative ranking made using each method.

proach has been used to obtain baseline quantitative estimates of N_2 fixed. The highest ^{15}N enrichment amongst several sugarcane varieties (i.e. the lowest fixing or possibly non N_2-fixing sugarcane variety) was selected as a reference to estimate the least possible proportion and amount of N_2 fixed that should be expected (M. Fried, personal communication). This approach could also be used for nitrogen fixing trees where it is difficult to select a suitable reference crop.

In the results shown in Table 1, using the lowest fixer, var. Chippewa inoculated with strain 'D' as such a control,

$$\% \text{ Ndfa in Amsoy} = \left(1 - \frac{0.0456}{0.1112}\right) \times 100 = 59\% \quad (3)$$

which compared well to the 65% obtained by using the assumed suitable non-nodulating soybean reference crop. Therefore, even in the absence of accurate quantitative information it is to be expected that, at least 59% of the N in Amsoy was derived from N_2 fixation, and also, that it was the highest N_2 fixer in this study. Valuable recommendations of agronomic significance can certainly be made from such determinations until accurate measurements can be made. Furthermore, it is quite likely that for many varieties of a given N_2 fixing species, several characteristics, e.g., rooting patterns, and N accumulation profiles may be fairly similar, even though N_2 fixation may vary from negligible to high amounts; screening such lines could, in addition, reveal some non-nodulating and potentially good reference crops. For species that nodulate, screening a large germplasm collection for lines that do not nodulate, or nodulate only when inoculated with certain strains of *Rhizobium* may produce some ideal or acceptable reference crops. For example, in attempts to identify non-N_2-fixing *Casuarina*, Dommergues (Personal Communication) planted several *Casuarina* branchlets in long trays of sand heavily inoculated with *Frankia*, and observed the seedlings for nodulation.

Obligatory requirement for reference crops

For the accurate quantification of N_2 fixed, a satisfactory reference crop is often needed. Even then, errors introduced through the mis-match of reference and fixing plant have frequently been overstressed. This is illustrated by examining the % Ndfa values and their associated errors in Fig. 1. These values were calculated by the isotope dilution equation (Fried and Middelboe, 1977; McAuliffe et al., 1958) using hypothetical ^{15}N enrichments, plus or minus 10% error in the ^{15}N enrichments absorbed by fictitious reference crops, and assuming a single ^{15}N enrichment ($0.2 \pm 0.02\%$ ^{15}N a.e.) in the fixing crop (Hardarson et al., 1988). The errors are high for low fixation rates, e.g., below 30% Ndfa, become smaller as fixation increases, and above 80% the errors are so small that it does not matter statistically whether the selected reference crop is the one that assessed for example, 1.3, 1.4 or 1.5% ^{15}N a.e. instead of an assumed accurate value of 1.6. The calculated standard errors from the mean at 33, 50, 80 and 87.5% Ndfa are approximately ± 13.7, 10.1, 4.01 and 2.5, respectively. Therefore, the precision of the methodology improves greatly as N_2 fixation increases. Reichardt et al. (1987) have provided experimental evidence in support of this hypothesis. Thus, in contrast to those legume species and varieties for which N_2 fixation has been shown to be consistently high, such as in many forage legumes in which % Ndfa is close to 90% or more (Haystead and Lowe, 1977; Henzell et al. 1968; Ledgard et al., 1985b; Vallis et al., 1977), or where R is more or less stable with time (Butler, 1987; Poth et al., 1986), this mismatch between reference and fixing crop would normally result in only small deviations from the true values of N_2 fixed (Hardarson et al., 1988). It is under situations where N_2 fixation is low, e.g., when plants are harvested early (before much N_2 has been fixed), and also, R decline is rapid, that the selection of a well-matched reference crop for quantifying N_2 fixed is most critical.

The same argument raised in discussing Figure 1 can be adduced for differences in soil A-values that have been reported, when different N rates and different reference crops have been used. The data in Table 2 are unpublished results of experiments conducted at the same time in a field in Egypt as part of an International FAO/IAEA Coordinated Research Programme on N_2

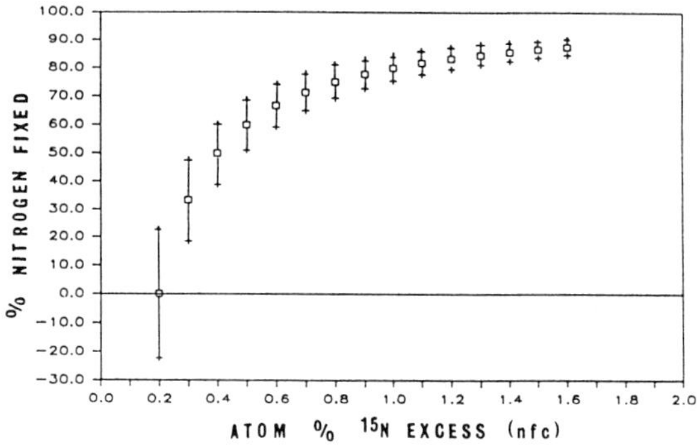

Fig. 1. Estimated values of % Ndfa in a hypothetical legume containing 0.2 atom % ^{15}N ±0.02 SD and a range of assumed atom % ^{15}N excess ±10% SD in potential nonfixing reference crop (nfc). The *bars* represent the upper and lower limits of standard deviation of the estimates (Hardarson et al., 1988).

fixation. Because N_2 fixation was high in fababean (high A-fixed value), the seemingly high differences in the soil A-values assessed by the different reference crops had little effect on the estimates of % Ndfa in fababean. These differences in soil A-values however resulted in significant differences in the estimates of % Ndfa in chickpea, in which N_2 fixation was lower.

Also, it is incorrect to assume that any differences in A-values measured by different reference crops will translate into absolute differences in N_2 fixed. This is, because, for N_2 fixation estimation, the A-value differences eventually have to be multiplied by a fertilizer use efficiency factor (E_f) of the legume (Fried and Broeshart, 1975), which is always less than unity, since in a soil, a plant never uses all the N applied as a fertilizer. These differences in A-values therefore translate into correspondingly lower N_2 fixed differences than is reflected by the A-value differences. From our experience, E_f values for many N_2-fixing legumes range less than 0.1 to 0.2 (or <10 to 20%), and are rarely greater than 40%. This explains why in a study of Wagner and Zapata (1982), although the soil A-value assessments differed much among the different reference crops (434 to 705 kg N ha^{-1}), the estimates of N_2 fixed did not differ much (ranged from 125 to 144 kg N ha^{-1}), because of the low E_f in fababean (6.1%). In a legume in which E_f is 5%, an A-value difference of as much as 60 between two reference crops would therefore mean a difference of only 3 kg N ha^{-1} of N_2 fixed, and not 60.

Reference crop selection

As discussed, the errors resulting from selecting a poor reference crop could be significant when N_2 fixation is low. Under such situations, it is crucial to obtain well-matched reference and fixing crops, or adopt practices that will lessen errors due to the mis-match.

Many studies that have quantified N_2 fixed may have been conducted without any preliminary evidence that the reference crops were suitable. This may be particularly so for N_2 fixing species and cultivars for which non-nodulating isolines exist. In such cases, it has often been assumed that these non-nodulating isolines are ideal reference crops for their fixing isolines (Fried et al., 1983). The evidence however does not indicate that this is always so (Rennie, 1979; 1982; Ruschel et al., 1979).

Fried et al. (1983) and Vose and Victoria (1986) have discussed the most important criteria that should be observed in selecting a suitable reference crop. It is however doubtful that enough attention has always been paid to these. We shall examine these briefly.

Table 2. Effect of differences in A values assessed by different controls on estimate of % N derived from atmosphere (% Ndfa) by fababean and chickpea (M R Hamissa, unpubl. data)

	A value (soil + fixed N) (kg N ha^{-1})	A fixed value as assessed using (kg N ha^{-1})				% Ndfa using control (%)				Range % Ndfa
		Barley (20)	Barley (100)	Wheat (20)	Wheat (100)	Barley (20)	Barley (100)	Wheat (20)	Wheat (100)	
Fababean	1719	1517	1516	1504	1543	87	87	86	89	86–89
Chickpea	295	94	92	79	120	30	29	25	38	25–38

Soil A values of reference crops: Barley (20) 202; Barley (100) 203; Wheat (20) 216; Wheat (100) 176, kg N ha^{-1}.

a) *Screening for absence of nitrogen-fixing activity in the reference crop*

Many non-legumes have most probably been used as reference crops on the assumption that they do not fix N_2. For many tropical C-4 grasses for example, this may be incorrect, since according to Vose and Victoria (1986), few of them do not support associative N_2 fixation. Using such plants as reference crops will therefore underestimate N_2 fixed in the test crop to the extent to which the reference crop itself obtained N from fixation (i.e. ignoring other sources of error). The impact of such errors on the accuracy of N_2 fixation measurements may however not be serious if N_2 fixed by the reference is low, while being high for the fixing crop. For example, in a study by Danso (unpublished), it was observed that although a few nodules developed late in the growth of the uninoculated soybean reference crop (as a result of rhizobial contamination from the inoculated treatments), the estimate of N_2 fixed using this reference crop was similar to that in which a non-nodulating isoline was used. Similar observations were reported by Cornet et al. (1985) and Gauthier et al. (1985). However, when N_2 fixed by the fixing crop is low, such errors could result in gross underestimates.

It has been suggested that it may be possible to use a N_2-fixing crop to which a high rate of inorganic N has been applied as a reference crop, on the assumption that a certain level of inorganic N may completely inhibit N_2 fixation (Rennie et al., 1978). For example, Vose and Victoria (1986) observed that the substantial N_2 fixation activity in Sudan grass at 20 kg N ha^{-1} was eliminated at 100 kg N ha^{-1}. Sudan grass at the higher but not the lower N rate could thus be used as a suitable reference crop (assuming it met other requirements). Sanginga et al. (1990) made a similar observation with N_2 fixation measurements in *Leucaena leucocephala* and *Acacia albida* using the corresponding uninoculated trees fertilized with 20 or 100 kg N ha^{-1} as reference plants. This therefore implies that the isotope dilution approach cannot be used under such situations, since the fixing crops would have to receive lower (to avoid significant inhibition of N_2 fixation) and therefore unidentical levels of inorganic N. In some legumes, however, N fertilizer rates, far in excess of those normally used could not completely inhibit N_2 fixation (Hardarson et al., 1991; Richards and Soper, 1979; Ruschel et al., 1982; Wagner and Zapata, 1982). Reliance on this approach without prior proof of complete suppression of N_2 fixation is, therefore, risky.

The acetylene reduction assay (Hardy et al., 1968) has frequently been used to screen potential reference crops for N_2-fixing ability (Boddey and Victoria, 1986; Phillips and Bennett, 1978; Ledgard et al., 1985b; Rennie and Dubetz, 1986). Also, the $^{15}N/^{14}N$ ratios in non-nodulated isolines, and uninoculated nodulating legumes in soil without native rhizobia, or those inoculated with ineffective strains have sometimes been compared with those in other potential reference crops grown on the same soil (Fried et al., 1983; Giller and Witty, 1987; Rennie, 1982); if similar, it is assumed that the potential reference crops are also non-N_2-fixers (Fried et al., 1983; Rennie et al., 1982). It is essential to examine uninoculated legumes for unexpected effective nodules from cross contamination, especially in field-grown plants (e.g. Cornet et al., 1985; Gauthier et al., 1985).

Although a crop may not fix N_2 when grown pure, it may derive N_2 fixed from a companion N_2-fixing plant (Broadbent et al., 1982; Van Kessel et al., 1985; Walker et al., 1956). By utilizing N, part of which was derived from N_2 fixation, these non-fixing crops may no longer be suitable as reference crops (Broadbent et al., 1982; McAuliffe et al., 1958). The transfer of fixed N_2 is still questionable, as several studies have not been able to demonstrate significant N transfer in field-grown plants (Butler, 1987; Danso et al., 1987; Ledgard et al., 1985c). In contrast, Broadbent et al. (1982) reported significant transfer of fixed N_2 in a mixed-pasture study, and therefore suggested that ^{15}N isotope dilution methods are not suitable for estimating N_2 fixed in grass-legume mixtures. There is, however, a need to examine critically the evidence for N transfer, since some of the results are not conclusive enough, or the methods used are questionable. One of these methods assumes that a lower $^{15}N/^{14}N$ ratio in a reference crop grown in mixture with a legume, compared to when grown pure, is a proof of N_2 transfer from the legume to the mixed reference crop

(Eaglesham et al., 1981; Fried et al., 1983; Vallis, et al., 1967). Based on this, the proportion of N derived from transfer of fixed N_2 in mixed cropping systems is calculated as follows (Hardarson et al., 1988):

$$1 - \frac{\%\ ^{15}\text{N a.e. in mixed crop}}{\%\ ^{15}\text{N a.e. in sole crop}} \qquad (4)$$

What has, however, not been usually established is, whether these plants, having been subjected to unidentical cropping situations, still maintained similar growth and N uptake patterns. If not, it is possible that the observed ^{15}N enrichment differences could be due to differences in N uptake patterns under the two different situations. In one of the few studies in which this was tested, Henzell et al. (1968) obtained similar growth of sole cropped siratro and siratro intercropped with Rhodes grass on a ^{15}N-labelled organic matter-amended soil. No fixed N_2 transfer was in that case demonstrated. On the other hand, Vallis et al. (1977) observed ^{15}N uptake by grass was greater in the presence of a legume, than in its absence. Also, it is not unusual to sometimes observe a higher ^{15}N enrichment in a mixed than in a sole reference crop (e.g., Hardarson et al., 1988), a reversed N transfer situation that may have possibly arisen from dissimilar N uptake patterns by plants in the two systems. The effect of cropping system on growth and N uptake patterns of plants therefore needs to be investigated further, as it could possibly alter some of the evidence for fixed N_2 transfer. Simple time-course studies of N uptake could be used to discern any significant differences in N uptake profiles between these reference crops (Henzell et al., 1968). This approach has been used by Boller (1988) to show that perennial ryegrass grown in mixture with white clover took up more N late in a growth cycle than in monoculture during the seeding year, and there was no consistent pattern between mixed vs monoculture Italian ryegrass.

Even if it is established conclusively that fixed N_2 is transferred from a fixing to a reference crop, it is necessary to ascertain when this has a significant influence on estimates of N_2 fixed. The use of this reference crop becomes a serious problem mainly when the level of N_2 fixation is low. When N_2 fixation is high, the introduced errors are usually small, as shown by Hardarson et al. (1988). Thus, although Hardarson et al. (1988) estimated that up to 16% of the N in the ryegrass in a mixed alfalfa sward could have been at one stage derived from N transfer, the sole reference crop estimated 83% Ndfa in alfalfa, compared to 81% by the mixed ryegrass reference crop. Similar observations have been reported by Vallis et al., (1977).

b) *Root distribution*

Although the reference and fixing crops are required to absorb soil N with the same R (Fried et al., 1983; Ledgard et al., 1985b; Witty, 1983), they are not required to have similar rooting depths, equal root density and distribution (Danso, 1985; Fried et al., 1983; Rennie, 1984; Rennie and Kemp, 1984; Vose and Victoria, 1986). They are only expected to have access to the same ratio of the applied ^{15}N label and soil N, and may therefore absorb N from anywhere within the soil profile into which the ^{15}N was incorporated, and not necessarily from exactly the same depth (Danso, 1985; Fried et al., 1983; Vose and Victoria, 1986). This should permit some variation in the root systems of reference and fixing crops, as illustrated in Figure 2. In this example, each of non-fixing crops 2 and 4, which differ in rooting lengths, sizes and/or distribution would be suitable as reference crop (assuming they meet other requirements) for estimating N_2 fixed by fixing crop 1, because they all obtain their N from the topsoil N pool (Zone A) within

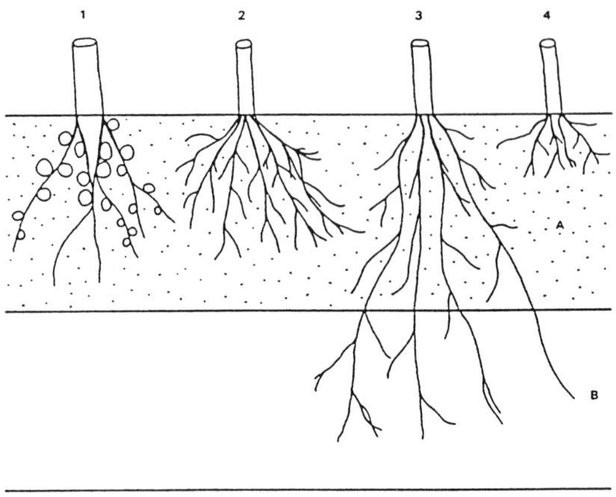

Fig. 2. Effect of root depth on ^{15}N/^{14}N uptake

which the ^{15}N was incorporated. Experimental verification of this has been provided by Ledgard et al. (1985c) who observed that ryegrass was a suitable reference for clover, even though ryegrass assimilated N from greater depths than clover. Furthermore, Ledgard et al. (1985c) reported that root activity and indigenous plant-available soil N were small in the 10–40 cm zone, relative to the 0–10 cm zone. The greater activity of ryegrass roots at depth therefore had no significant effect on R or the determination of N_2 fixed by clover. Fried et al. (1983) similarly suggested that most annual crops would normally absorb N from the top horizon or plough layer (about 0–30 cm). Even deep-rooted trees have been shown to obtain most of their nutrients from the topmost 30 cm soil (Anon., 1975; Mahendrappa et al., 1986). Thus, although crop 3 in Fig. 2 may not theoretically qualify as a suitable reference crop, should the available N or root activity be very low in the B horizon, it would not be surprising if the estimated value of N_2 fixed did not deviate significantly from that made by using a better-matched reference crop. This may explain why reference crops with different rooting patterns have been shown to be equally good for estimating N_2 fixed in several legumes (Fried et al., 1983; Rennie, 1982; Rennie and Kemp, 1984; Wagner and Zapata, 1982; Witty, 1983).

A practical problem is, how to ensure a uniform vertical distribution of the applied fertilizer. Should the vertical distribution of ^{15}N vary widely, the ^{15}N enrichments in several reference crops that differ greatly in depth of N uptake are bound to vary, as shown by Boddey and Victoria (1986). Attempts to achieve fairly homogeneous ^{15}N distribution in soil include watering or irrigating immediately after fertilizer application (Jensen, 1986; Rennie, 1982; Rennie and Kemp, 1984; Vallis et al., 1977; Vasilas and Ham, 1984), or the physical incorporation of the ^{15}N fertilizer to the desired soil depth within which nutrient uptake is presumably most pronounced (Boddey et al., 1984; Eaglesham et al., 1981; Patterson and LaRue, 1983; Talbott et al., 1982; Witty, 1983). According to Rennie (1986), however, choosing the most appropriate reference crop is likely to be more important than the problem of non-uniform mixing of the applied ^{15}N, on the assumption that any inherent errors would be similar for both fixing and reference crop and thus of reduced consequence. More research is needed to quantify the influence of root length differences and ^{15}N enrichment gradients in agricultural soils on estimates of N_2 fixed.

The simplest method for examining rooting characteristics is to carefully dig a trench to expose the roots, or to uproot plants for root measurements (Fried et al., 1983). The depth to which roots grow may, however, not indicate correctly the zone of active nutrient uptake. A more precise approach involves injecting ^{32}P or ^{33}P-labelled solutions, or both, at various soil depths and distances from the base of the plant. Differences in the activities of ^{32}P or the ratio of $^{32}P/^{33}P$ with depth then reflect the differences in root activity at the various depths and distances (Anon., 1975). Nitrogen fertilizers, labelled with ^{15}N have also been used to determine root-activity patterns of legumes and reference crops (Ledgard et al., 1985a; Steele and Littler, 1987). Nitrogen-15 is non-radioactive and therefore its use is not hazardous. However, the high solubility and mobility of inorganic N within the soil profile can complicate the interpretation of such point-placement studies.

c) *Relative N uptake patterns*

Dissimilarity in relative N-uptake patterns of the reference and fixing crops in a soil with a high R decline-rate constant (Witty, 1983) is an important factor in the controversy surrounding the accuracy of ^{15}N estimates of N_2 fixed. The major problem is related to the unavoidable and often rapid decline of R with time following conventional ^{15}N fertilizer application to soil (Fried et al., 1983; Jensen, 1986; Witty, 1983).

In soils in which R is practically stable, the accurate estimation of N_2 fixed is not affected by dissimilar N uptake profiles of fixing and non-fixing crops (Witty, 1983). This is illustrated by situation ‘A’ in Figure 3; irrespective of when N uptake begins or terminates in this case, the final R assessed by the reference crop would be 0.3% ^{15}N a.e. However, the more realistic situation to expect is that of ‘B’, when conventional ^{15}N fertilizer is added to most field soils. The rapid R decline is attributable to N cycling

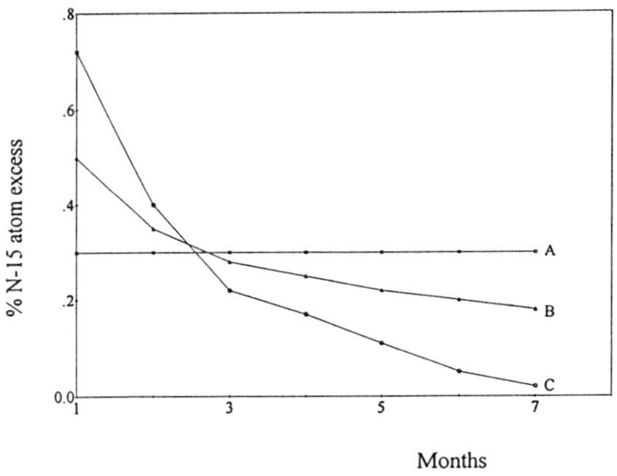

Fig. 3. Hypothetical decline of ^{15}N atom excess with time.

process (plant uptake, N losses, transformation into less-available forms, mineralization of unlabelled soil N, etc.). The final R assessed will, therefore, vary for different crops which absorb N in different patterns, and a suitable reference crop and N_2-fixing crop must be fairly matched in N-uptake profiles over the entire growth period, to obtain accurate estimates (Fried et al., 1983; Witty, 1983). Ill-matched N-uptake profiles in one study resulted in a 50% error in the N_2 fixation estimate (Witty and Ritz, 1984). Under such circumstances therefore, procedures that resulted in less fluctuation in ^{15}N/^{14}N ratio of soil with time (e.g., from situation 'C' to 'B' in Fig. 3) significantly reduced the errors (Witty and Ritz, 1984).

Several approaches have been used to assess that various crops are well matched in their N-uptake patterns. The simplest involves determining total N in sequential harvests, which is very useful for selecting suitable reference crops, or at least, for discarding obvious mis-matches (e.g., Jensen, 1986). Using this approach to match various reference and fixing crops Witty (1983) estimated that differences in N uptake patterns during the early growth cycle introduce greater errors than when these differences occur later; the higher R-decline rate immediately following ^{15}N application, combined with the greater initial availability of ^{15}N than late during growth may account for this.

Boller (pers. commun.) has used a different approach to compare N uptake patterns of different crops. This is done by comparing the N yield increment within each growth interval as percentages of the final N yield (rather than using the cumulative N). The average deviations in N yield increments between legume and reference crops over the whole season is then calculated, and used to assess the magnitude of any error. Using this approach, Boller found frequent and large differences in the N uptake patterns of *Trifolium repens* and *T. pratense* and the pair, *T. repens-Lolium perenne* were better matched than a *T. pratense-L. multiflorum* combination. However, because % Ndfa was high, these differences in N uptake profiles resulted in only small differences in the % Ndfa measurements, again demonstrating that reference crop selection is less critical in measurement of N_2 fixation in high N_2-fixing crops.

Comparisons of the ^{15}N enrichments of N_2 fixing plants when in non-fixing mode with those of other species have been used to assess the suitability (and thus similarity in N uptake pattern) of potential reference crops (e.g., Fried et al., 1983; Rennie, 1982; Wagner and Zapata, 1982). Fried et al. (1983) suggested that a non-nodulating legume isoline when available, is an ideal reference crop for its nodulating isoline on the assumption that these should be genetically similar in most aspects other than in the ability to nodulate and fix N_2. This is supported by many reported satisfactory estimates using non-nodulating isolines as reference crops (Boddey et al., 1984; Vasilas and Ham, 1984; Wagner and Zapata, 1982; Witty and Ritz, 1984). Contradictory observations have however been made by Patterson and LaRue (1983) and Rennie (1982). Rennie (1982) assessed the fixing crop in a non-fixing mode (e.g., uninoculated or inoculated with an ineffective strain in *Rhizobium*-free soil) to be more suitable than the non-nodulating isoline.

Double-labelling with ^{15}N and ^{35}S isotopes [($^{15}NH_4)_2$ $^{35}SO_4$] was used by Wagner and Zapata (1982) to indirectly assess the R values estimated by the fixing and several non-fixing crops. The rationale behind this is, that the estimated A-values of sulphur for both fixing and reference crops are based on absorption from the substrate only, and should therefore be similar for all crops that absorbed nutrients in a similar pattern and from the same pool. With this approach, Wagner and Zapata (1982) identified oil radish to be an unsuitable reference crop for fababean. Instead, one may just compare the ratio % Sdff/% Sdfs between the reference and fixing plant and use those with same ratio as the fixing plant.

Ledgard et al. (1985a) have developed a mathematical formula for calculating R assimilated by both fixing and reference crops. This involves growing N_2-fixing legumes and reference plants in soils receiving (1) none and (2) same fertilizer and N rate but of two or more widely differing ^{15}N-enrichments. The proportions of added and indigenous N assimilated by the plants were determined from the intercepts and slopes of regression lines, and from this, reference crops with similar N uptake profiles as the N_2-fixing legume *T. subterraneum* were identified. This method has however not been widely adopted.

These various attempts indicate the serious efforts that are being made to ensure that the reference crop does not hinder the accurate determination of N_2 fixation in crops. The relative N-uptake profiles or suitability of reference crops may however change under different climatic or ecological conditions (Fried et al., 1983; Papastylianou and Danso, 1989), and the adoption or rejection of a reference crop based on studies in other locations is therefore not recommended. This is illustrated by the study of Rennie and Dubetz (1986) in which they found no significant difference in values of N_2 fixed in a field in Vauxhall, using either barley or wheat as a reference crop. However, at another location, Lethbridge, barley gave significantly higher estimates of N_2 fixed in 7 out of 14 comparisons. The selection of an appropriate reference crop was therefore shown experimentally to be site specific, and may explain why Rennie (1982) found a non-nodulating soybean to be a poor reference crop for its nodulating isoline, despite its reported suitability in several studies conducted elsewhere (e.g. Boddey et al., 1984; Fried et al., 1983; Witty and Ritz, 1984). Giller and Witty (1987) suggested that differences in estimates of N_2 fixation they obtained in similar experiments replicated at different sites may have been confounded by the variable response of the reference crop to differences in climate and soil. Papastylianou and Danso (1989) observed that seasonal differences in rainfall affected the suitability of oats as reference crop for vetch. It therefore seems inappropriate to make general recommendations on the suitability of reference crops based on a one season or location trial. The reference crop should match the fixing crop under ideal conditions, and as suggested by Fried et al. (1983), also when stress conditions set in during the growth period.

^{15}N labelling and plant harvest dates

With R in a ^{15}N-fertilized soil usually declining with time, plants may sample dissimilar R unless the ^{15}N fertilizer is applied to the fixing and reference crop at the same time. Similarly, plants should not be harvested on different dates (Fried et al., 1983). The N_2 fixation estimates in those studies in which reference and fixing crops were fertilized on different dates or harvested at different periods are open to question.

Rennie and Rennie (1983) suggested that the validity of the ^{15}N soil labelling methodology should not require similarity in maturity times of fixing and reference crops. There is, however, a good case for the reference and fixing crops to be similar in growth development and maturity, rather than being harvested only at similar times. This is, because, the relative N accumulation rates (and therefore N uptake profiles) by crops are often linked to the growth stages. For example, in the study of Witty (1983), although several reference crops were harvested at the same time, the relative N uptake patterns differed (especially during the late growth stages) and closely corresponded with days to maturity. Thus, the rates of N uptake by oilseed rape, field beans and french beans declined earlier than those for clover and grass, in line with earliness of attaining maturity. As a result, oilseed rape in the study of Witty (1983) appeared to be a suitable reference crop for field beans and french beans. In contrast, N uptake by the pasture crops, clover and grass was still increasing at this period and as a result, oilseed rape was an unsuitable reference crop for clover, while grass, which matured later was. Maize, in another study by Witty and Ritz (1984), matured later than nodulating and non-nodulating soybean, and continued to increase in N content, in contrast to the soybeans. Consequently, maize, unlike non-nodulating soybean was found to be an unsuitable reference crop, giving several negative estimates of N_2 fixed, even though all crops were harvested at the same time.

The problem of unmatched maturity dates can be quite critical in comparisons of the N_2 fixing abilities of different varieties. In such instances, relative N_2 fixation assessments based on the % ^{15}N a.e. in plants could be misleading when the days to maturity differ significantly. It is essential to limit such ^{15}N comparisons to genotypes of similar maturity groups. This is illustrated by the study of Ruschel et al. (1982), in which the earliest-harvested variety (60 DAP) on the basis of relative % ^{15}N a.e. was ranked as the poorest N_2 fixer. Such a comparison is at best doubtful since those fixing crops harvested 30 days later continued to absorb soil N of continuously declining ^{15}N enrichment, and could eventually have contained lower ^{15}N enrichments than the earlier-harvested plants even if they were not better fixers. Having several replicates of a given reference crop and harvesting some plots along with each legume harvest may not always be a satisfactory solution as suggested by the study of Patterson and LaRue (1983) in which the N accumulation rates of the earlier maturing soybean varieties (cv. Wilkins and Clay) declined towards harvest at 97 DAP, contrary to the continued increase in N uptake of the later maturing non-nodulating Harosoy. Harvesting the non-nodulating soybean along with the earlier maturing fixing legumes in such a case would not solve the problem of an inappropriate reference crop. Maturity date differences in varieties are common, and therefore an important practical problem in N_2 fixation comparisons. It is advisable that when several varieties or species are to be compared, at least two or three different reference crops, possibly of differing maturing dates be grown together with the fixing crops to increase the probability of having an appropriate reference crop for fixing crops harvested on each date.

^{15}N application to soil and procedures

Several ^{15}N fertilizer formulations, rates, ^{15}N enrichments and methods of application have been used. Although it may be tempting to view this lack of standardization as a weakness of the ^{15}N techniques, differences in soils, crops, environmental conditions, analytical capabilities, financial resources, etc. would preclude such a general recommendation. An exhaustive coverage of the literature on ^{15}N application procedures has been provided by Chalk (1985). The N rate to use and method of ^{15}N incorporation into soil will be influenced by several factors, including the rate of N release, which may also be related with the type of N fertilizer used.

Release rates

A high or low rate of ^{15}N release may be desirable, depending on the objective. For ill-matched reference and fixing crops, errors in N_2 fixation estimates increase as the decline-rate constant of R increases (Witty, 1983; Witty and Ritz, 1984). Thus, by using slow-release ^{15}N-labelled fertilizers, which result in reduced decline in R (i.e., more stable R), errors due to mis-matched N-uptake patterns of reference and fixing crops are reduced (Witty and Ritz, 1984). However, although the slow release of N reduces the chances of fertilizer N being high enough at any one time to significantly suppress N_2 fixation in the fixing crop, it may not provide enough N for the growth of the reference crop. Some fertilizers designated as slow-release formulations have been tried, including gypsum-pelleted KNO_3, oxamide (Witty and Ritz, 1984) and isobutylidine diethyl urea (S. Danso and F. Zapata, unpublished). The ^{15}N-tagged forms of slow-release fertilizers, however, are not yet available commercially. Rennie (1986) and Rennie and Dubetz (1986) simulated a slow-release fertilizer formulation by mixing $(^{15}NH_4)_2SO_4$ with a nitrification retarding agent, N-serve (2-chloro-6-(trichloro-methyl)-pyridine), to successfully estimate N_2 fixed in field bean, chickpea, lentil, fababean and fieldpea. This is a simple approach, which is worth trying by others.

The need for a stable soil R is less important

when the reference and fixing crops are well matched and/or when N_2 fixation is very high. Under such circumstances, the advantages gained by using slow-release N fertilizers may be so small that conventional ^{15}N labelled fertilizers could be preferable, since they are less expensive and available commercially.

Ionic forms

The N in the ^{15}N-labelled fertilizer is, in most cases, added only as a tracer for N_2 fixation studies. Thus, unless the fertilizer form has a significant effect on N_2 fixation, or is unequally available to the reference and fixing crop, the ionic form used is of little importance. The tracer ^{15}N does not, therefore, always need to be the same as that normally used in fertilizer or agronomic trials in the locality (Danso et al. 1988). Where necessary, it should be possible to use two different fertilizer formulations in the same study, one for monitoring the agronomic effect, and a cheaper or sometimes the sole available ^{15}N form for tracing the sources of N in the legume (Danso et al., 1988).

There have however been some reported problems and discrepancies in using different ^{15}N labelled fertilizers to estimate N_2 fixed, in particular, those containing either the NH_4^+ or NO_3^- ionic forms, and these have been discussed by Chalk (1985). These discrepancies and problems are largely attributable to differences in (1) solubility, (2) movement of the labelled N within the soil profile, and (3) preferential utilization of different ionic forms by the different crops.

The solubility of a fertilizer form influences N availability to the plant. Nitrate forms were reported to result in greater N utilization efficiencies than NH_4^+ – containing fertilizers (Rennie, 1986; Vallis et al., 1967). Chalk (1985) suggested that unequal N release rates could give rise to differences in changes in R with time. These differences should however be of little importance in well-matched plants; the essential requirement is, the accurate estimation of R for each circumstance, rather than R being the same for different situations. This explains why Vallis et al. (1967) observed that although *Stylosanthes humilis* utilized 36% of the applied $(NH_4)_2SO_4$ compared to 72% of KNO_3, both ^{15}N labels produced similar % Ndfa values (91 and 90%, respectively). This can be explained by the corresponding differences in utilization of NH_4^+ and NO_3^- (47 and 71%, respectively) by the reference crop, *Chloris guyana*. In contrast, for reasons which are not clear, Ruschel and Saito (1981) detected N_2 fixation in *Phaseolus* using $^{15}NO_3^-$ form, but none when $^{15}NH_4^+$ was used.

The greater N uptake from highly soluble forms may enhance ^{15}N enrichment in plants, and consequently improve the precision of N_2 fixation estimates, although high solubility leading to enhanced N uptake could inhibit N_2 fixation, unless low quantities are applied. On the other hand, in coarse textured or irrigated soils, substantial leaching of highly soluble N forms may occur, resulting in reduced N availability. This was observed in one of the FAO/IAEA coordinated research experiments in Senegal (F. Ganry, personal communication) in which heavy rainfall followed almost immediately after the application of different levels of ^{15}N fertilizer. Because the soil was sandy, much leaching occurred, reducing all rates of N applied to almost identical low levels; the differential effect of N rates on N_2 fixation was not therefore observed, although differences in tolerance to N which varied with varieties had been observed in the previous year. It would be wrong to conclude in such a case that fertilizer N had no effect on N_2 fixation in any of the varieties tested.

The fertilizer form also influences N movement within the soil profile. The NH_4^- ion, for example, is less mobile than NO_3^-, particularly, in clayey or high organic matter soils. Thus, although the ^{15}N isotope techniques do not usually stress on equal root lengths of reference and fixing crops (Fried et al., 1983), the applied ^{15}N could, under some situations, become more accessible to some crops than others (e.g., as a result of movement into the subsoil). The R estimated by the reference crop in such a case may therefore not be valid for estimating N_2 fixed in the fixing crop. This was the situation when Ledgard et al. (1985b), observed a lower R for clover than for ryegrass when $K^{15}NO_3$ was washed into soil with 10 mm water. However, R did not differ when these same plants were irrigated with less (2 mm) water. At the higher watering rate, the deeper-rooted ryegrass obtained more of the ^{15}N which had moved into the subsoil, than the shallower-rooted clover. These differences resulted in a significantly higher % Ndfa estimate with the 10 mm irrigation (because of higher ^{15}N in the reference crop) than under the 2 mm water regime in which NO_3^- movement was more restricted. Under such experimental conditions, it is difficult to assess the effect of water levels per se on the measured N_2 fixation, distinct from possible methodological limitations. Under these circumstances, there is an unavoidable need for matched root length of reference and fixing crop, or for labelling procedures that ensure less drastic vertical gradients in ^{15}N. A further example is provided by the study of Steele and Littler (1987), in which higher N_2 fixation was estimated in clover that received $Na^{15}NO_3$ rather than $(^{15}NH_4)_2SO_4$. The data of $Na^{15}NO_3$ injection at various soil depths suggest that the reference crop had deeper roots than the fixing crop, and deeper injection of $Na^{15}NO_3$ resulted in higher % Ndfa estimates than shallow injection (Steele and Littler, 1987). The observed difference between the N_2 fixation estimates using the two fertilizers was therefore probably due to the slower movement of NH_4^+ into deeper layers compared to NO_3^-. In perennial crops however, these effects may be transient, as shown in the later harvests made by Ledgard et al. (1985b), in which the earlier differences in N_2 fixation disappeared with time. Possible reasons for this could be the increased N_2 fixation with time, as well as the differences in root lengths (and also R) becoming smaller with time.

Some plants may differ in their preference for NH_4^+ and NO_3^- forms, while the rates at which NH_4^+ is converted to NO_3^- could in some cases differ in the rhizospheres of different plants (R. Bradley, personal communication). Should these differences occur under the rhizospheres of N_2 fixing and non-fixing reference crop, they would then not be estimating the same R. Thus, although Vallis et al. (1967) obtained similar % Ndfa values using NH_4^+ and NO_3^- forms, there is the need for more work on the selective absorption of NH_4^+ and NO_3^- by different plants grown on the same soil to find out whether this is widespread or not, and if this significantly influences N_2 fixation measurements in other localities.

Methods of ^{15}N application

The great majority of N_2 fixation studies have been conducted with the ^{15}N fertilizer sprayed as a water solution onto the soil surface (Hardarson et al., 1984; Jensen, 1986; McAuliffe et al., 1958; Phillips et al., 1983; Vallis et al., 1977; Witty, 1983). In few cases, the solution was injected directly into the root zone, using special applicator syringes (Rennie, 1986; Rennie and Dubetz, 1986; Steele and Littler, 1987; West and Wedin, 1985). Although the data of Fried and Broeshart (1981) and Fried et al. (1983) indicated that the method of ^{15}N application did not significantly influence N_2 fixation estimates in fababeans, applying as a solution allows small quantities of ^{15}N fertilizer to be dissolved in large volumes of water, thereby facilitating a more uniform spatial distribution than by spreading small quantities of solid material (Fried et al., 1983; Vose and Victoria, 1986). Also, except under the special circumstances of wind drift reported by Rennie (1986), we do not see any special advantages in the injection technique over the much simpler surface spray application. In contrast, we expect the better distribution of ^{15}N from the solution spray should result in less errors than by the injection technique.

Time of ^{15}N application

The ^{15}N fertilizer has frequently been applied at sowing (McAuliffe et al., 1958; Phillips et al., 1983; Rennie and Kemp, 1984). In some cases, it has been allowed to equilibrate in the soil for some time before planting (Patterson and LaRue, 1983) or has been delayed until soon after germination (Hardarson et al., 1984a; Jensen, 1986; Ruschel et al., 1982; Vasilas and Ham, 1984). Delayed planting after ^{15}N application allows time for much of the applied ^{15}N to be incorporated into organic matter fractions (Freitas et al., 1984), with subsequent slow release and decreased errors arising from mis-matched N-uptake patterns of reference versus fixing crop. Post-emergence application of ^{15}N on the other hand allows for greater planning flexibility. In this case, the ^{15}N fertilizer addition can be stopped or postponed if germination problems, etc. are noticed (Duque et al., 1985; Rennie, 1986). Besides, where necessary, thinning of plants can be performed before the ^{15}N addition (e.g., Ruschel et al., 1982). In addition, one can verify that both reference and fixing crops are well-matched in germination and initial growth (and thus possibly in N uptake) before the ^{15}N is applied. In one case, one of the crops had to be planted a week earlier to ensure matched growth, but the ^{15}N was applied to both crops at the same time (Ledgard et al., 1985). Any possible N accumulation and N_2 fixation within approximately one week after sowing is bound to be negligible, and unlikely to lead to serious errors in the final estimates of N_2 fixed due to the delayed ^{15}N application.

Rates of N application

Although for most N_2 fixation studies, ^{15}N fertilizers have been used mainly as tracers, the fact that the N they contain can affect growth and N_2 fixation should not be overlooked in selecting the N rate to use. In many studies, N rates may have been used without prior knowledge of how they would affect N_2 fixation under the local conditions. For the use of the ^{15}N techniques to estimate N_2 fixation, a zero ^{15}N fertilizer rate (which would have been ideal) is only possible with the ^{15}N natural abundance method. As a general rule the ^{15}N applied must be high enough to ensure satisfactory ^{15}N detection in plants, without reducing N_2 fixation. Low N rates will therefore necessitate the use of highly enriched ^{15}N fertilizers (e.g. West and Wedin, 1985). However, cost considerations may not always make this approach the best (Rennie, 1986; Danso et al., 1987). For example, although both 2 g N of 50% ^{15}N and 100 g N of 1% ^{15}N urea supply 1 g of ^{15}N, the former is usually more expensive, because of more laborious processes in the manufacture. Increasing the quantity of ^{15}N fertilizer (if it does not significantly inhibit N_2 fixation) with a corresponding decrease in ^{15}N enrichment is thus economically more advisable (Danso et al., 1988). In any case, appropriate N rates would normally differ from soil to soil and for different fixing crops. These are however seldom higher than 20 to 30 kg N ha^{-1} for the legume.

Our experience with grain legumes in the IAEA Seibersdorf Laboratory and with International FAO/IAEA Co-ordinated Research Programmes has shown that 0.1 g ^{15}N m^{-2} (equivalent to 10 or 20 kg N ha^{-1} of 10% and 5% ^{15}N a.e., respectively) results in satisfactory levels of ^{15}N enrichment in legume tissues. Generally, average % ^{15}N a.e. values at physiological maturity for the several experiments were around 0.15 to 0.20 for soybean, 0.25 to 0.30 for common bean, 0.10 to 0.15 for chickpea, 0.2 to 0.25 for cowpea and 0.06 to 0.10 for fababeans. As expected, these values were significantly lower than those of the non-fixing reference crops. These ^{15}N enrichments in the legumes are satisfactory for the analytical capability of all mass spectrometers, and these ^{15}N rates could even be reduced to a third or half. However, for emission spectrometry using Jasco, Sopra, NOI-4 or NOI-5 instruments, higher ^{15}N rates (around 0.2 to 0.3 g ^{15}N m^{-2}) may be more satisfactory (especially for high N_2 fixers like fababeans) because of their lower ^{15}N detection threshold than mass spectrometers. The new emission spectrometers, NOI-6e (supplier: Fischer Analysen Instrumente GmbH, Leipzig, Germany), SOPRA GS-5 (supplier: SOPRA Co., Bois-Colombes, France) however have greatly improved sensitivity of ^{15}N detection of close to natural abundance, compared to the minimum of about 0.5% ^{15}N abundance in the older models (Table 3). The suggested application of fertilizers enriched with 0.5 to 1% ^{15}N a.e. (Rennie, 1986) is therefore rather low (unless at N levels inhibitory to N_2 fixation) for emission spectrometry detection of ^{15}N in plant samples, but adequate for mass spectrometry.

High versus low N rates may be added to the reference and fixing crops respectively, to estimate N_2 fixed. This approach overcomes the limitation of poor growth of the reference crop when soil N is inadequate, while avoiding the suppression of N_2 fixation by high inorganic N (Fried and Broeshart, 1975). In some situations a higher N level for the reference crop may even

Table 3. Comparison of ^{15}N analytical methods (Axmann, 1990)

Instrument and method	Emission spectrometer		Mass spectrometer	
	'old' type[a]	'new' type = NOI-6e[b]	with an inlet system	in connection with an automatic nitrogen analyzer ANA
Material which can be used for analyses	Only Kjeldahl extracts (NH_4^+ – salts)	Only Kjeldahl extracts (NH_4^+ – salts)	Only Kjeldahl extracts (NH_3^+ – salts)	Any type of material (dry combustion)
Lowest amount of N required for the analysis	2 μg (0.2)	10 μg	50 μg	50 μg
Lowest at. % ^{15}N which can be measured	0.55 (0.45)	Near to natural abundance 0.36	Even depleted less than 0.36	Even depleted – less than 0.36
Rel. standard deviation	~ ± 2–3%	~ ± 0.6–1%	± 0.02%	± 0.2%
Total time for analyses in min.	30–40 +5 for Kjeldahl	2 +5 for Kjeldahl	10–30 +5 for Kjeldahl	6
Time for measurement only, in min.	5	2	10–30	4
Time for sample-preparation in min.	25–35 +5 for Kjeldahl	0 +5 for Kjeldahl	0 +5 for Kjeldahl	2 (for weighing)
Cost of equipment in US$	~35.000	~45.000–50.000	~120.000–200.000	~130.000

[a] Only sealed discharge tubes can be used for measurements.
[b] No preparation of discharge tubes is needed; the instrument has a flow through discharge tube; N_2 gas is generated from Kjeldahl extract and Sodiumhypobromide solution.

prevent premature senescence, ensure better matched N uptake and growth patterns (Patterson and LaRue, 1983), or even completely inhibit N_2 fixation in an otherwise N_2-fixing reference crop (Sanginga et al., 1990; Vose and Victoria, 1986). The use of differential N rates normally necessitates an estimation of soil A-value, because the direct isotope dilution equation of McAuliffe et al. (1958) cannot be applied in this case to estimate N_2 fixed in plants grown on soils which by virtue of the different amounts and enrichments of ^{15}N added are no longer similar. It should though be mentioned here that % Ndfa can be quantified by one equation even when different N rates are applied to the fixing and non-fixing crops (Hardarson and Danso, see this volume). This same equation can be used for the isotope dilution method also.

The A-value approach relies on the assumption that the A-value of soil remains constant at different N rates (Fried and Dean, 1952). The validity of this assumption has been criticized

(Broadbent, 1970; Hamilton et al., 1992), although some studies have shown experimentally that it can be relied upon (Legg and Stanford, 1967). Duhoux and Dommergues, (1985); Gauthier et al. (1985) and Vasilas and Ham (1984), for example, obtained similar N_2 fixation values using either the isotope dilution methodology, or the A-value approach. Because the pioneering report of Fried and Broeshart (1975) used 20 and 100 kg N ha^{-1} for the fixing and reference crop respectively, it is significant that most N_2 fixation studies that have used the A-value approach have adopted these rates. There is, however, no reason why such divergent rates should always be used. For example, although Deibert et al. (1979) reported dissimilar A-values as N rates changed, the differences were greatest as the gap between the N rates increased. It appears that as N rates increase beyond a certain high level (depending on the soil type, etc.), the soil A-value may not always remain constant (Hardarson et al., 1991). Also, Henzell et al. (1964) showed that the uptake of N was a linear function of the NH_4NO_3 applied up to about 200 kg N ha^{-1} (a requirement for constant A-value), and fell significantly at 400 kg N ha^{-1}. It is likely that at supraoptimal N rates, more N is applied than can be effectively utilized by the plant, and the postulated proportional increase in N uptake at higher N levels (upon which the A-value concept hinges) then no longer holds (Hardarson et al., 1991). Since the objective of higher N rate addition to the reference crop is to ensure healthy plant growth rather than to ensure maximum growth, a rate as low as would allow normal or healthy plant development, e.g., to avoid the physiological growth problems induced by excessive N deficiency (Boddey et al., 1984; Harper, 1974; Vasilas and Ham, 1984; Witty and Ritz, 1984) is all that is necessary for the reference crop. Thus, in a legume study in which 20 to 30 kg N ha^{-1} does not significantly affect N_2 fixation, and about 40 kg N ha^{-1} ensures good growth of the reference crop, the adoption of 20 kg N ha^{-1} to the fixer and 40 kg N ha^{-1} for the reference crop (a 1:2 ratio) would be more preferable than the highly contrasting rates of 20 and 100 kg N ha^{-1} (1:5 ratio) frequently used.

It is difficult to assess the extent to which human error may have contributed to the A-value controversy, although this could be significant. Clearly, the A-value approach is more demanding in practice than the isotope dilution approach. For example, great care is needed when applying vastly different N rates to field crops, since small systematic errors in the fertilizer application may translate into significant errors at the lower N rate, but not at the higher rate.

The practice of applying the ^{15}N fertilizer sometime ahead of planting may be stretching the limits of the A-value assumptions too far. It is quite probable that the ratio between the initially applied two or more N rates would change after a long period of incubation. The assumption that the two or more crops are sampling fertilizer N in a certain applied ratio may then not hold. Where the A-value approach is to be used, it is advisable to plant immediately after ^{15}N addition, or to apply the ^{15}N after planting.

Frequency of ^{15}N application

The one-time application of labelled N, at ^{15}N enrichments satisfactory for N_2 fixation measurements over more than one season (e.g. for perennial fixing crops), may be very expensive and/or inhibitory to N_2 fixation. Besides, the greatest problem with this approach, even with annuals is the sharp decline in $^{15}N/^{14}N$ ratio of soil, often necessitating the rigorous selection of reference crop. To overcome these problems, equal-sized, multiple additions of small amounts of ^{15}N fertilizer during various times of the growth cycle were first used by Vallis et al. (1967), to estimate N_2 fixed by Townsville stylosanthes. Since then, several studies using this approach have been reported (Boddey et al., 1983a and b; Danso et al., 1988; Edmeades and Goh, 1978; Vallis et al., 1977). Boddey et al. (1984) found that the frequent additions of small doses of ^{15}N fertilizer ensured high ^{15}N fertilizer enrichments in soil, and resulted in small declines in R, compared to the single application. Frequent ^{15}N additions may therefore assist in minimizing errors due to mis-match between reference and fixing crop. With this repeated ^{15}N application on a plot, the estimate of % N derived from the recent ^{15}N fertilizer addition is confounded by the unknown contribution from the residual ^{15}N. Boddey et al. (1984) however reported that N_2 fixation values obtained from organic matter-labelled plots were similar to those that received seven sequential applications of $(^{15}NH_4)_2SO_4$. These results indicate that the effect of frequent, small ^{15}N fertilizer additions on estimates of N_2 fixed, if any, were rather small, and may be of little consequence especially at high levels of fixation. Danso et al. (1988) and Labandera et al. (1988) suggested that ^{15}N uptake from residual N in soil is often insignificant compared to the substantial N uptake from recent applications, for which reason they obtained satisfactory estimates from such sequential ^{15}N additions. Based on field studies, Steele and Littler (1987) have similarly suggested that estimates of N_2 fixed, following multiple application of ^{15}N on plots appear to be mostly influenced by the most recent addition of ^{15}N.

Rennie (1985, 1986) has, however, criticised the use of repeated ^{15}N additions on plots. The arguments and evidence he adduced in support of his claim appear unconvincing. For example, it is not clear how the suggested differences in rhizosphere N and thus ^{15}N enrichments of the fixing and reference crops would cause more serious errors when ^{15}N application is repeated on the same plots than when ^{15}N is applied once, at planting. On the contrary, the consecutive application of N of higher ^{15}N enrichment than that present in the soil could obliterate differences in soil R that are likely to develop with time under the rhizospheres of the fixing and reference crops, where the ^{15}N was applied only at planting (Danso et al., 1987; Labandera et al., 1988). In addition, a single N rate, in order not to inhibit N_2 fixation and yet be detected over several seasons, may have to be highly enriched (i.e. costly), and may also necessitate the use of expensive and highly sophisticated analytical capability that may not be available in many laboratories.

Immobilized N

Nitrogen when in organic forms is released slowly (dictated by the rate of mineralization) for plant uptake. This gradual release of mineral N from ^{15}N labelled organic material results in a more stable R compared with inorganic ^{15}N fertilizer (Butler, 1987; Fried et al., 1983; Ladd et al., 1981; Legg and Sloger, 1975; Witty and Ritz, 1984). For a 12 week period, for example, the ^{15}N enrichment of the mineral N in soil into which ^{15}N-labelled organic matter had been incorporated remained almost constant (Butler and Ladd, 1985). Boddey et al. (1984) also reported an initial decline in the $^{15}N/^{14}N$ ratio in available soil N in an organic matter amended soil, after which it remained fairly constant, while Ladd et al. (1981) showed that the R in a soil into which ^{15}N-labelled medic was added did not change appreciably over a 7-month period, and that the ^{15}N enrichment measured in the plant was similar to that of the soil extract. A perfectly stable R in theory, but difficult to achieve in practice as shown by the data of Boddey et al. (1984) and Papastylianou and Danso (1991) precludes the need for reference crops, since the $^{15}N/^{14}N$ ratio of the soil extract, at any point in time would reflect what was absorbed by the fixing crop (e.g., Butler, 1987; Poth et al., 1986). Any decrease in plant $^{15}N/^{14}N$ ratio compared to that of the extract is then attributable to the N_2 fixed from the atmosphere (Chalk et al., 1983; La Favre and Focht, 1983).

Because of problems associated with ^{15}N decline, i.e. need for selecting appropriate reference crop, it is tempting to advocate the general adoption of organic matter labelling approaches. There are however a few drawbacks, some of which have been discussed by Chalk (1985). Although it may be true that a perfectly stable R may, in most cases, be difficult to achieve, what is most important is, that the incorporation of ^{15}N-labelled organic matter in soil results in improved stability of R compared to using inorganic ^{15}N fertilizers, so as to decrease errors in N_2 fixation estimates due to mis-matched N uptake patterns of reference and fixing crops (Witty and Ritz, 1984). Although Chalk (1985) mentioned the difficulty in ensuring complete equilibration of ^{15}N and soil N, this should not always be a target. The realistic objective is to significantly reduce the rapid decline in R, common in many soils fertilized with conventional ^{15}N-labelled fertilizers, and this objective has certainly been achieved in several studies in which it cannot be claimed that complete equilibration of ^{15}N with soil organic matter had taken place (Boddey et al., 1984; Butler and Ladd, 1985; Ladd et al., 1981). For example, in the study of Boddey et al. (1984), the ^{15}N enrichment in non-nodulating soybean grown on soil enriched with ^{15}N labelled organic matter fell from 0.56 on 29 DAP to 0.37 at 138 DAP, a slower decline than would have been observed with conventional ^{15}N fertilizers (Fried et al., 1983). Furthermore, most of the decline (to 0.4) occurred during the first 92 DAP. During the latter 46-day period therefore, it is reasonable to assume that the $^{15}N/^{14}N$ ratio in such a soil was operationally stable.

Compared to using conventional chemical fertilizers labelled with ^{15}N, a longer waiting period is needed with the organic matter approach to allow reasonable equilibration with soil N to take place (Fried et al., 1983; Rennie, 1986). However, the ^{15}N enrichments in such soils also remain high for longer periods (Fried et al., 1983), and may thus be used over several years without the need for new ^{15}N additions, in so far as treatments that drastically introduce differences in soil chemical/physical properties on different portions of the soil are not used. When ^{15}N-labelled plant materials are available from previous experiments, these could be incorporated without extra cost for ^{15}N, and this would, in addition, cut down on the time needed to produce ^{15}N-labelled plant tissue.

To reduce the waiting period, some approaches that accelerate the conversion of the applied ^{15}N into the organic matter pool have been adopted. These involve the addition of high energy substrates, such as sucrose and glucose simultaneously with the ^{15}N-labelled fertilizer or plant residue (Fried et al., 1983; Legg and Sloger, 1975; Ronis et al., 1985; Talbott et al., 1982; Wagner and Zapata, 1982; Witty and Ritz, 1984). The subsequent increased microbial proliferation accounts for the accelerated immobilization of the added ^{15}N. It is, however, uncertain as to what direct or indirect effects this expected change in microbial population or activity and changes in other soil properties would have on N_2 fixation estimates (Fried et al., 1983).

What is however certain is, that estimates of N_2 fixed are generally higher when ^{15}N-labelled organic matter is incorporated into soil than when other procedures are used (Chalk, 1985). Thus, most N_2 fixation estimates in soybean that fall in the range of 80% and higher Ndfa have been obtained in soils supplemented with organic matter (e.g., Boddey et al., 1974; Henson and Heichel, 1984; Legg and Sloger, 1975; Patterson and LaRue, 1983; Wagner and Zapata, 1982). This is in contrast with the commonly reported average of around 50% (Rennie et al., 1982). The increased % Ndfa values are most likely due to the enhanced microbial immobilization of not only the added ^{15}N, but also, much of the available soil N (Henzell et al., 1968; Ladd et al., 1983; Wagner and Zapata, 1982; Witty and Ritz, 1984). In principle, the amounts of organic materials that have been added have usually been excessive, and drastic changes in available soil N resulting in N deficiency in the reference crop, and in artificially enhanced N_2 fixation in the fixing crop are therefore to be expected. As shown by Rennie (1986), the use of lower organic matter rates (equivalent to N application rates of only 9 to 13 kg ha^{-1}) than usually reported, resulted in similar N_2 fixation estimates in field beans as without these residues. Undoubtedly, the rationale behind the incorporation of high rates of organic matter is to ensure the release of significant levels of ^{15}N above natural abundance, in order to avoid analytical problems. For example, Rennie (1986) detected only 0.0028 to 0.0078% ^{15}N a.e. in the reference plants and 0.0016 to 0.0034% ^{15}N a.e. in the bean plants. These values, which are more usefully expressed in δ^{15}N units need highly expensive and precise double collector mass spectrometers for detection. The ^{15}N values would, however, have been much higher if the initial ^{15}N enrichments in these organic materials had also been higher. For example, when Boddey et al. (1984) incorporated approximately 11 tons of sorghum ha^{-1} containing 4.018% ^{15}N a.e. into soil (in contrast to the 0.6123 to 0.9509% ^{15}N a.e. in the organic materials used by Rennie, 1986), the non-nodulated soybean grown on this soil contained 0.4% ^{15}N atom excess at 138 DAP. With the high initial ^{15}N enrichment, Boddey et al. (1984) could have lowered the organic matter rates (e.g., by 1/3 to 1/2) and still detected significant ^{15}N enrichments above ^{15}N natural abundance. Data showing significant release of ^{15}N from crop residues (Ladd et al., 1981) support the suggestion that lower rates of organic matter than currently used may be satisfactory for N_2 fixation studies. After the addition of 57 kg N ha^{-1} of 3.6% ^{15}N a.e. labelled organic matter (equivalent to approximately 2,000 kg of organic matter ha^{-1}) to soil and 8 months incubation, wheat plants grown for 7 months on this soil contained as much as 0.458% ^{15}N a.e. (Ladd et al., 1981). Also, the proportions of ^{15}N labelled and unlabelled N in the mineral N pool in this study did not change appreciably during the cropping period.

Another cause for the organic matter induced increase in % Ndfa is the high C/N ratio of the ^{15}N-labelled plant residues (e.g. Henson and Heichel, 1984; Henzell et al., 1968; Patterson and LaRue, 1983; Talbott et al., 1982; Wagner and Zapata, 1982). Such materials would result in considerable soil N immobilization, and the use of plant residues of lower C/N ratios approaching those of the non-stable organic matter fractions in the soil should minimize soil N immobilization rates (Papastylianou and Danso, 1991). Also, a low C/N ratio will ensure rapid mineralization and will enable lower N rates to be used. Thus, efforts should be made to incorporate organic materials whose C/N ratios are only slightly higher than those of the soils being used. Also, in some studies, the incubation periods have been rather short (Legg and Sloger, 1975; Wagner and Zapata, 1982; Witty and Ritz, 1984), and appear unsatisfactory. Longer incubation periods would have assisted in better ^{15}N equilibration with soil N and would reduce any side effects on available soil N. This is illustrated by the rapid decline that occurred in R in ^{15}N organic matter labelled soils in which the plants were grown after only short or no period of incubation (Boddey et al., 1984; Legg and Sloger, 1975), in contrast to those in which practically stable R was obtained after longer waiting periods of incubation (e.g., Butler, 1987; Poth et al., 1986).

It is also true that it is difficult in practice to obtain a uniform distribution of the labelled organic matter in soil (Chalk, 1985). According to Rennie (1986) however, the most essential aspect should be to ensure a state of fairly operationally uniform mixing to minimize errors due to non-homogeneous distribution of ^{15}N.

Although it is the general tendency to accept soils labelled with ^{15}N from organic matter as the most promising and giving the most reliable estimates of N_2 fixed, from the above discussions, it seems that this may not be so yet, and that more research is needed, for example, on the appropriate rates and ^{15}N enrichments to use. In the FAO/IAEA Programme at Seibersdorf Laboratory, we are continuing with studies on the rototilling of young, succulent, ^{15}N labelled plants into the very soil on which they had been grown (Fried et al., 1983), and have obtained additional information (unpublished data) that show that when used properly, this approach does not enhance N_2 fixation. Instead of using highly lignified material, the plant to be rototilled is ploughed in when succulent (low C/N ratio, closer to that in soil than when mature), and another crop is planted on the same plot, but without further ^{15}N addition, to be again rototilled before maturity. To ensure high ^{15}N in the plant material (i.e., high ^{15}N utilization), the ^{15}N applied to the first crop is provided in split applications, so that the highest application is made when roots are fully developed, and N uptake is likely to be greatest (Afza et al., 1987). Early heavy N fertilization when the roots are not fully developed and N demand is low (Afza et al., 1987) would result in lowered utilization of the ^{15}N fertilizer, and therefore organic material with lower ^{15}N enrichment.

One of the greatest limitations with the organic matter labelling approach is, that under N deficient conditions, the growth of the reference crop suffers, and in extreme cases, may not even develop to full maturity. This problem will become even greater if the crops grown are perennials which should be harvested periodically over long durations. Furthermore, in attempting to use a soil labelled with organic material for more than one season (a postulated advantage), the N status of the soil and its effect on the reference crop should not be overlooked.

Residual ^{15}N in soil

Soil nitrogen balance studies have revealed that much of the N applied in fertilizers usually remains in the soil after crop harvest (anonymous, 1975; Rao and Shinde, 1984). Most of this N is however converted into various organic forms (Chalk, 1985) and mineralized slowly for plant uptake. The residual ^{15}N from previous experiments, has therefore been used for estimating N_2 fixed in plants (Danso et al., 1988; Kohl and Shearer, 1981; Rennie, 1982; Rennie et al., 1982; Steele and Littler, 1987). Several sites that have been used previously for ^{15}N-aided field experiments and which could be usefully employed for N_2 fixation studies exist. As suggested by Kohl and Shearer (1981), differences between the $^{15}N/^{14}N$ ratios of such soils and the atmosphere are often sufficiently large that isotopic fractionation should not cause serious errors in the analysis and interpretation of the data. In their study, a soil that had received 168 kg N ha^{-1} of a fertilizer enriched with 9.5% ^{15}N a.e. contained 106.9 $\delta\,^{15}N$ units (>0.04% ^{15}N a.e.) 12 1/2 years later and was therefore good for measuring N_2 fixation. Also, Steele and Littler (1987) quoted an unpublished study in which ryegrass was grown on soil 498 days after ^{15}N labelled $(^{15}NH_4)_2SO_4$ was applied, and yet found about 300 $\delta^{15}N$ units in the herbage. The rate of decline in R in soils that have previously received ^{15}N fertilizers is very low, compared to new ^{15}N additions. For example, the study of Kohl and Shearer (1981) showed that after 51 days of plant growth, the ^{15}N enrichment in the reference crop had fallen only slightly from 106.9 δ units to 104.5. Similar results have been reported from several other studies (e.g., Focht and Stolzy, 1978; Poth et al., 1986). Depending on how constant R remains, it has been possible to measure N_2 fixation without the need for reference crops (La Favre and Focht, 1983; Poth et al., 1986). The fraction of N_2 fixed in pigeon pea was calculated by La Favre and Focht (1983) using the following equation:

$$1 - \frac{\%\ ^{15}N \text{ a.e. in plant}}{\%\ ^{15}N \text{ a.e. in soil}} \qquad (5)$$

The ease and practicality of employing residual ^{15}N for estimating N_2 fixation depends on the extent to which ^{15}N eventually gets incorporated into various fractions of soil organic matter, and the subsequent rate of N mineralization. In very sandy soils, residual N following ^{15}N application is expected to be small, while in heavy textured or high organic matter soils with high microbial activity, residual ^{15}N could be substantial.

The sensitivity of the ^{15}N measuring equipment available dictates the feasibility of using residual ^{15}N for N_2 fixation estimations. Errors considered to be negligible, e.g., in the $^{15}N/^{14}N$ analysis of highly enriched samples can be very serious at very low levels of ^{15}N enrichments, necessitating the use of high precision dual collector mass spectrometers which may not be available in most laboratories. On the other hand, by applying 10 kg N ha^{-1} of 13.4% ^{15}N a.e. $(NH_4)_2SO_4$ (equivalent to about 0.13 g ^{15}N m^{-2}) to soil, this resulted in plots containing appreciably high levels of ^{15}N (lowest, 0.0365% ^{15}N a.e.) and was successfully used to estimate N_2 fixed in two consecutive years without further ^{15}N incorporation (Fried et al., 1983). In the second year the ^{15}N values in the Fried et al. (1983) study, although low for most emission spectrometry determinations, were high enough for all mass spectrometers.

The non-uniform distribution of residual ^{15}N in the soil N cited by Chalk (1985) needs to be examined further. However, the CV of 11.9% in N_2 fixation estimates made in the year of ^{15}N application, compared with the 9.3% when ^{15}N uptake was from the residual ^{15}N in soil (Fried et al., 1983) and the low CVs that have been reported elsewhere for N_2 fixation determinations using residual ^{15}N (Ledgard et al., 1985; Rennie, 1982; Rennie et al., 1982) suggest that this may not be a serious problem.

Serious problems with the use of residual ^{15}N could however arise when there are substantial differences in root and litter turnover of fixing perennial and reference crops (Danso et al., 1988, 1992; Labandera et al., 1988). Such differences would result in lower $^{15}N/^{14}N$ in the rhizosphere of the fixing crop (because of the lower enrichment of the atmospheric N_2 that is fixed), compared to the reference crop, and would result in the fixing and non-fixing plants absorbing soil N of different R. A basic concept of the ^{15}N methodology for measuring N_2 fixation is thus violated. Also, Phillips et al. (1983) reported that ^{15}N in the herbage of the harvest that was not immediately preceded by ^{15}N application was largely the result of N turnover and retranslocation from roots and crowns to new shoot growth. The large differences that frequently exist in the ^{15}N enrichments of the fixing and reference crop may therefore introduce serious errors when the residual ^{15}N approach is used, unless suitable corrections are made.

Sampling of plant materials

The ^{15}N enrichments in various plant parts (e.g. seed, herbage, crowns and roots) frequently differ (Butler, 1987; Butler and Ladd, 1985; Fried et al., 1983; Jensen, 1986; Ladd, 1981; Rennie et al., 1978). Measurements of % Ndfa, based on only one plant part, e.g., seed (Ruschel et al., 1982) may therefore not adequately represent N_2 fixed in the whole plant (Ladd, 1981; Phillips et al., 1983a). In basing these estimations on individual plant organs, it is being assumed that the ^{15}N distribution and N translocation patterns in individual organs of the fixing versus reference crop are similar throughout growth. This assumption may not hold under many situations. On the other hand, plant organs differ significantly in densities, and it is therefore difficult to obtain a representative single sample comprising all plant parts to be used for % N and ^{15}N determination. The ^{15}N differences in plant parts, arise largely from differences in the absorption rate and translocation of N of different isotopic composition to the different plant parts at the various growth stages (Fried et al., 1983) and have been cited as constituting a problem in the ^{15}N isotope methods for measuring N_2 fixation (Chalk, 1985; Heichel et al., 1981; 1984; Jensen, 1986). This need not, however, pose any serious problems, since it may be overcome or minimized by stripping, sampling, and where appropriate, subsampling each of these recognizable plant parts (which would be expected to have fairly uniform N isotopic composition) separately for % N and % ^{15}N a.e. analyses (Fried et al., 1983). The following equations are then used for calculating a weighted average of % ^{15}N a.e. (WAE) for the whole plant, and subsequently % Ndfa.

$$\text{WAE} = \frac{\text{AE(a)} \times \text{TN(a)} + \text{AE(b)} \times \text{TN(b)} + \text{AE(c)} \times \text{TN(c) etc}}{\text{TN(a + b + c) etc.}} \tag{6}$$

where a, b, c, etc. represent various plant parts, AE is % ^{15}N atom excess, and TN is total nitrogen

and % Ndfa (whole plant)

$$= \left(1 - \frac{\text{WAE (fc)}}{\text{WAE (nfc)}}\right) 100 \tag{7}$$

This is a standard accepted procedure adopted in most N_2 fixation studies that use ^{15}N. The major disadvantage with splitting plants into different parts for analyses is, the increase in the number of samples, and therefore, increased costs and manpower time. However, in many cases, only pods (with enclosed seeds) may need to be analysed separately from other above ground-parts. A more recent approach adopted in the IAEA Laboratory in Seibersdorf (Danso et al., 1992) involves sampling the individual parts at harvest, drying and grinding the separate parts and then mixing the powder according to the ratios of their respective dry matter yields to get a reliable composite sample.

There may even be situations when whole plants or individual plant parts (of the reference crop in particular) could be used with justification, thus cutting down on labour and analytical costs. For example, when N_2 fixation was high, Ledgard et al. (1985) measured similar % Ndfa in the whole clover plant or in shoot alone, although there were differences in ^{15}N enrichments in the different organs of clover as well as the reference crop. Results obtained by Zapata et al. (1987a) also showed that while ^{15}N enrichments in pods and straw of fababean or N_2-fixing soybean differed significantly, they were similar in either non-nodulated soybean or barley at maturity. Haystead and Lowe (1977) and Heichel et al., (1981) have also reported fairly similar ^{15}N enrichments in shoots, roots and stubble of white clover and alfalfa, respectively. Where uniformity in ^{15}N enrichment has been demonstrated (e.g., from preliminary studies), there is no need to split plants into separate organs for analyses. It has to be recognised though that similarity in ^{15}N enrichment in different organs of a particular crop may not necessarily hold under all circumstances. Also, in many forage legumes, the problem of non-uniform distribution in the above-ground parts may not be as serious as in grain legumes, since forage legumes are normally harvested before significant seed production.

Estimates of N_2 fixation have mostly been made only on the above-ground herbage (Hardarson et al., 1984; Danso et al., 1987; Phillips and Bennett, 1978; Williams et al., 1977; Zapata et al., 1987). This is based on the knowledge that most of the fixed N_2 in grain legumes is translocated to the above ground parts (Danso and Kumarasinghe, 1990). The measurements thus exclude any N_2 fixed in roots or incorporated into the soil (Poth et al., 1986). In many grain legumes, roots or underground structures may constitute less than 10% of the standing crop biomass (Deibert et al., 1979; Minchin et al., 1978). Therefore where only small differences in ^{15}N enrichment exist between the below and above-ground parts, the errors made by excluding roots in total N_2 fixed measurements will most likely be less than 10% (see Danso and Kumarasinghe, 1990). This is in contrast to many forage legumes and trees for which significant proportions of their total dry matter and N content are below-ground (Heichel et al., 1981; 1984; Sanginga, 1990). Sanginga et al. (1990) found that roots of *L. leucocephala* contained about 60% of the N_2 fixed in the whole plant, and serious underestimation of N_2 fixation would thus result by ignoring the roots. In addition, where significant differences in ^{15}N isotopic composition of various plant parts exist, N_2 fixation may be overestimated by ignoring the usually higher $^{15}N/^{14}N$ in the roots. The necessity to measure the N_2 fixed in roots would however depend on the relative distribution of ^{15}N in the organs of fixing crops being compared, and the question being asked, e.g., the proportion for N_2 fixed in the harvested material, distinct from the amount fixed in the plant, or in the soil/plant system. Since most plant N and fixed N_2 are in the seed (Vasillas and Ham, 1985; Zapata et al., 1987a, b), N_2 fixed in seed is often closest to that fixed in the whole plant (Vasillas and Ham, 1985). When compelled to (e.g., in case of severe leaf shedding), seeds could therefore be the most reliable organ to use for relative comparisons of N_2 fixed. In contrast, where total N balance or the impact of crop growth on soil fertility is at issue, the errors have greater impact and the amount of N_2 fixed in roots has to be measured.

Witty and Ritz (1984) and Jensen (1986) have shown the magnitude of errors that can result from calculations of N_2 fixed when the unlabelled seed N is not corrected for. These effects are however likely to be most pronounced at early harvests, but given the normally small contribution from seed N (seldom more than $5\,\mathrm{kg\,N\,ha^{-1}}$) to the total N in most grain and forage legumes, it is doubtful that this would result in significant errors in most other situations, particularly, where N_2 fixation at maturity is high. Besides, the correction for seed N of Witty and Ritz (1984) assumed 100% translocation of the stored N in seed to the harvested plant parts, an assumption that cannot be completely justified, as shown by the studies of Jensen (1986). However, in perennial forage and tree crops in which substantial amounts of N may be stored in below-ground parts and stubble or stumps, ignoring stored N may result in high errors. The organic N reserves in these organs (of different ^{15}N enrichment than soil N) are retranslocated to the new foliage, upon which much of the initial plant growth depends (Phillips et al., 1983b). This carryover N could be corrected for at any harvest, using the following equation:

$$AE_{(2-1)} = \frac{AE_2 \times N_2 - AE_1 \times N_1}{N_2 - N_1} \tag{8}$$

where

$AE_{(2-1)}$: is the ^{15}N enrichment of the N absorbed during the growing period,

$AE_{(2)}$: is the measured ^{15}N enrichment of whole plant at harvest or time 2.

and

$N_{(1)}$: is the N in the roots and/or stubble at time 1 (or before ^{15}N was applied).

This equation could also be used for correcting for seed N, but as indicated, whole plants have to be assayed since the equation assumes 100% translocation of carryover N if even only above-ground parts are harvested.

Precision

Although the high CVs reported in some studies (e.g. Danso et al., 1987; Fried et al., 1983) sometimes lead to questions on the precision of ^{15}N isotopic methods, several comparative studies have shown that the ^{15}N methods are superior to other techniques (Gauthier et al. 1985; Hardarson et al., 1984b; Rennie, 1984; Ruschel et al., 1979; Talbott et al., 1982). The seemingly high CVs that have sometimes made it difficult to discern statistical differences in ^{15}N-derived estimates of N_2 fixed in different treatments may sometimes be real (for example, when crops are grown in a very heterogeneous soil) rather than due to the methodology. The extent to which soil heterogeneity can lower the precision of ^{15}N-derived estimates of N_2 fixation has been shown experimentally by Reichardt et al. (1987). It is unlikely that any other method would have given better precision than the ^{15}N methodology under such highly variable soil conditions. The CVs of 1 to 5% obtained by Labandera et al. (1988) and Danso et al. (1988) in field experiments and 1.3% by Rennie (1982) under controlled conditions are better than can be expected for most field methods. Indeed few studies on the sensitivity of methodologies for measuring N_2 fixation have addressed the real problems associated with soil heterogeneity. Nevertheless, several approaches have been shown to minimize soil heterogeneity problems. These include the use of (many) small rather than large (and few) isotope subplots (Broadbent et al., 1980), and the pairing of fixing and reference crops on adjacent plots or close to each other (Fried et al., 1983; Reichardt et al., 1987; Talbot et al., 1983).

Although selecting homogeneous sites improves the precision of N_2 fixation measurements, it is doubtful if this is always the best approach, because it may not reflect accurately what is occurring in the field. The agronomist, for example, may be more interested in the quantification of N_2 fixed over the whole field rather than on only a selected homogeneous segment, which avoids the natural variation in the field. It, therefore, seems inappropriate to always ascribe high CVs in N_2 fixation to errors or weakness in the ^{15}N methodology, since they may reflect real differences in N_2 fixation along various sites. In such highly variable situations, even though decisions based on sound statistical principles may be difficult to make, rational agronomic decisions which would yield highly fruitful results could still be made.

Plant to plant variation may sometimes contribute greatly to poor precision, although this has hardly received the attention it deserves. The seriousness of this has been demonstrated in the study of Dohoux and Dommergues (1985) and Gauthier et al. (1985) in which the high variability they observed for N_2 fixation estimates in *Casuarina* was attributed largely to tree to tree variation. Duhoux and Dommergues (1985) observed that based on tree size variation, average % Ndfa varied from 16 (in the smallest trees) to 70 (in the biggest-sized trees).

Conclusion

The ^{15}N isotopic methods for measuring N_2 fixation are not perfect. The major difficulty has been in deciding on the appropriate reference plant to use. They are still the most reliable methods for measuring N_2 fixation and should provide near accurate results when used carefully. There is evidence that these methods have provided good data upon which sound management recommendations have been made. In our opinion the most serious problems with the ^{15}N methods are their inappropriate use and misconceptions. It is hoped that this article will assist in remedying some of these problems.

References

Afza R, Hardarson G and Zapata F 1987 Effects of delayed soil and foliar N fertilization on yield and N_2 fixation of soybean. Plant and Soil 97, 361–368.

Amarger N, Mariotti A, Mariotti F, Dorr J C, Bourguignon

C and Lagacheri B 1979 Estimate of symbiotically-fixed nitrogen in field-grown soybeans using radiations in the nitrogen-15 natural abundance. Plant and Soil 52, 269–280.

Anon, 1975 Root activity patterns of some tree crops. Results of a Five-year Co-ordinated Research Programme of the Joint FAO/IAEA Division of Atomic Energy in Food and Agriculture. STI/DOC/10/170. IAEA, Vienna.

Axmann H 1990 Methods for ^{15}N determination. *In* Use of Nuclear Techniques in Studies of Soil-Plant Relationships. Ed. G Hardarson. pp 55-59. IAEA, Vienna.

Boddey R M and Victoria R L 1986 Estimation of biological nitrogen fixation associated with *Brachiaria* and *Paspalum* grasses using ^{15}N labelled organic matter and fertilizer. Plant and Soil 90, 265–292.

Boddey R M, Chalk P M, Victoria R L and Matsui E 1983a The ^{15}N isotope dilution technique applied to the estimation of biological nitrogen fixation associated with *Paspalum notatum* cv. batatais in the field. Soil Biol. Biochem. 15, 25–32.

Boddey R M, Chalk P M, Victoria R L, Matsui E and Döbereiner J 1983b. The use of the ^{15}N isotope dilution technique to estimate the contribution of associated biological nitrogen of *Paspalum notatum* cv. batatais. Can. J. Microbiol. 29, 1036–1045.

Boddey R M, Chalk P M, Victoria R L and Matsui E 1984 Nitrogen fixation by nodulated soybean under tropical field conditions estimated by the ^{15}N dilution technique. Soil Biol. Biochem. 16, 583–588.

Boller B C and Nösberger J 1988 Influence of dissimilarities in temporal and spatial N-uptake patterns on ^{15}N-based estimates of fixation and transfer of N in ryegrass-clover mixtures. Plant and Soil 112, 167–175.

Bremner J M 1977 Use of nitrogen-tracer techniques for research on nitrogen fixation. *In* Biol. Nitrogen Fixation Farming Systems of the Tropics. Eds. A Ayanaba and P J Dart. pp 335–352. Wiley, Chichester, UK.

Broadbent F E 1970 Variables affecting A-values as a measure of soil nitrogen availability. Soil Science 110, 19–23.

Broadbent F E, Nakashima T and Chang G Y 1982 Estimation of nitrogen fixation by isotope dilution in field and greenhouse experiments. Agron. J. 74, 625–628.

Burris R H 1974 Biological nitrogen fixation. Plant Physiol. 54, 443–449.

Butler J H A 1987 The effect of defoliation on growth and N_2 fixation by *Medicago* sp grown alone or with ryegrass. Soil Biol. Biochem. 19, 273–279.

Butler J H A and Ladd J N 1985 Growth and nitrogen fixation by *Medicago littoralis* in pot experiments: Effect of plant density and competition from *Lolium multiflorum*. Soil Biol. Biochem. 17, 355–361.

Butler J H A and Ladd J N 1985 Symbiotically-fixed and soil-derived nitrogen in legumes grown in pots in soils with different amounts of available nitrate. Soil Biol. Biochem. 17, 47–55.

Chalk P M 1985 Estimation of N_2 fixation by isotope dilution: An appraisal of techniques involving ^{15}N enrichment and their application. Soil Biol. Biochem. 17, 389–410.

Chalk P M, Douglas L A and Buchanan S A 1983 Use of ^{15}N enrichment of soil mineralizable N as a reference for isotope dilution measurements of biologically fixed nitrogen. Can. J. Microbiol. 29, 1046–1052.

Clair, St. D A, Wolyn D J, DuBois J, Burris R H and Bliss F A 1988 Field comparison of dinitrogen fixation determined with nitrogen-depleted and nitrogen-15 enriched ammonium sulphate in selected inbred backcross lines of common bean. Crop Sci. 28, 773–778.

Cornet F, Otto C, Rinaudo CT, Diem H G and Dommergues Y 1985 Nitrogen fixation by *Acacia holosericea* grown in field simulated conditions. Acta Oecol. 6, 211–218.

Danso S K A 1985 Methods of estimating biological nitrogen fixation. *In* Biological Nitrogen Fixation in Africa. Eds. Ssali H and Keya S O. pp 213–244. MIRGEN, Nairobi.

Danso S K A and Kumarasinghe K S 1990 Assessment of potential sources of error in nitrogen fixation measurements by the nitrogen-15 isotope dilution technique. Plant and Soil 125, 87–93.

Danso S K A, Hardarson G and Zapata F 1986 Assessment of dinitrogen fixation potentials of forage legumes with ^{15}N technique. *In* Proceedings of a Workshop on Potentials of Forage Legumes in Farming Systems of Sub-Saharan Africa. Eds. I Haque, S Jutzi and P J H Neate. pp 26–57. ILCA, Addis Ababa, Ethiopia.

Danso S K A 1986 Review: Estimation of N_2-fixation by isotope dilution: An appraisal of techniques involving ^{15}N enrichment and their application comments. Soil Biol. Biochem. 18, 243–244.

Danso S K A, Zapata F, Hardarson G and Fried M 1987 Nitrogen fixation in fababeans as affected by plant population density in sole or intercropping systems with barley. Soil Biol. Biochem. 19, 411–415.

Danso S K A, Labandera C, Pastorini D and Curbelo S 1988 Nitrogen fixation in a two-year old white clover-*Fescue* pasture: Influence of nitrogen fertilization. Soil Biol. Biochem. 20, 261–262.

Danso S K A, Hardarson G and Zapata F 1988 Dinitrogen fixation estimates in alfalfa-ryegrass swards using different nitrogen-15 labelling methods. Crop Sci. 28, 106–110.

Deibert E J, Bijerigo M and Olson R A 1979 Utilization of nitrogen-15 fertilizer by nodulating and non-nodulating soybean (*Glycine max.*) isolines. Agron. J. 71, 719–723.

Duhoux E and Dommergues Y 1985 The use of nitrogen fixing trees in forestry and soil restoration in the tropics. *In* Biological Nitrogen Fixation in Africa. Eds. Ssali H and Keya S O. pp 384–400. MIRGEN, Nairobi.

Duque F F, Neves M C P, Franco R L and Boddey R M 1985 The response of field grown *Phaseolus vulgaris* to *Rhizobium* inoculation and the quantification of N_2 fixation using ^{15}N. Plant and Soil 88, 333–342.

Eaglesham A R J, Ayanaba A, Rao V R and Eskew D L 1981 Improving the nitrogen nutrition of maize by intercropping with cow-pea Soil Biol. Biochem. 13, 169–171.

Edmeades D C and Goh K M 1978 Symbiotic nitrogen fixation in a sequence of pastures of increasing age measured by a ^{15}N dilution technique. N.Z. J. Agric. Res. 21, 623–628.

Focht D D and Stolzy L H 1978 Long-term denitrification studies in soils fertilized with $(^{15}NH_4)_2SO_4$. Soil Sci. Soc. Am. J. 42, 894–898.

Freitas de J R 1984 Estimation of dinitrogen-fixation by sugar cane, *Saccharium* sp. and soybean, *Glycine max.*, grown in soil with nitrogen-15-labelled organic matter. Plant and Soil 82, 257–261.

Fried M and Broeshart H 1981 A further extension of the method for independently measuring the amount of nitrogen fixed by a legume crop. Plant and Soil 62, 331–336.

Fried M and Broeshart H 1975 An independent measurement of the amount of nitrogen fixed by a legume crop. Plant and Soil 43, 707–711.

Fried M, Danso S K A and Zapata F 1983 The methodology of measurement of N_2 fixation by non-legumes as inferred from field experiments with legumes. Can. J. Microbiol. 29, 1053–1062.

Fried M and Dean L A 1952 A concept concerning the measurement of available soil nutrients. Soil Sci. 73, 263–271.

Fried M and Middelboe V 1977 Measurement of amount of nitrogen fixed by legume crop. Plant and Soil 43, 713–715.

Gauthier D, Diem H G, Dommergues Y R and Ganry F 1985 Assessment of N_2 fixation by *Casuarina equisetifolia* inoculated with *Frankia* ORSO21001 using ^{15}N methods. Soil Biol. Biochem. 17, 375–380.

Giller K E and Witty J F 1987 Immobilized ^{15}N-fertilizer sources improve the accuracy of field estimates of N_2-fixation by isotope dilution. Soil Biol. Biochem. 19, 459–463.

Hamilton S D, Smith C J, Chalk P M and Hopmans P 1992 A model based on measurement of soil and plant ^{15}N enrichment to estimate N_2 fixation by soybean (*Glycine max* L. Merrill) grown in pots. Soil Biol. Biochem. 24, 71–78.

Hardarson G, Danso S K A, Zapata F and Reichardt K 1991 Measurements of nitrogen fixation in fababean at different N fertilizer rates using the ^{15}N isotope dilution and "A-value" methods. Plant and Soil 131, 161–168.

Hardarson G, Zapata F and Danso S K A 1984a Effect of plant genotype and nitrogen fertilizer on symbiotic nitrogen fixation by soybean cultivars. Plant and Soil 82, 397–405.

Hardarson G, Zapata F and Danso S K A 1984b Field evaluation of symbiotic nitrogen fixation by rhizobial strains using ^{15}N methodology. Plant and Soil 82, 369–375.

Hardarson G, Zapata F, Danso S K A and Golbs M 1988 Effect of N fertilization on N_2 fixation in fababean (*Vicia faba* L. var. minor). *In* Nitrogen Fixation: Hundred Years After. Eds. G Bothe, F J de Brujin and W E Newton. Gustav Fischer, Stuttgart. 814 p.

Hardarson G, Zapata F and Danso S K A 1988 Dinitrogen fixation measurements in alfalfa ryegrass swards using nitrogen-15 and influence of the reference crop. Crop Sci. 28, 101–105.

Hardy R W F, Burns R C and Holster R D 1973 Applications of the acetylene-ethylene assay for measurement of nitrogen fixation. Soil Biol. Biochem. 5, 47–48.

Hardy R W F, Holsten R D, Jackson E K and Burns R C 1968 The acetylene-ethylene assay for N_2 fixation: Laboratory and field evaluation. Plant Physiol. 43, 1185–1204.

Harper J E 1974 Soil and symbiotic nitrogen requirements for optimum soybean production. Crop Sci. 14, 255–260.

Haystead A and Lowe A G 1977 Nitrogen fixation by white clover in hill pasture. J. Br. Grassl. Soc. 32, 57–64.

Heichel G H, Barnes D K and Vance C P 1981a Nitrogen fixation of alfalfa (*Medicago sativa*) in the seeding year. Crop Sci. 21, 330–335.

Heichel G H, Barnes D K and Vance C P 1981b Evaluating elite alfalfa lines for N_2-fixation under field conditions. *In* Genetic Engineering of Symbiotic Nitrogen Fixation and Conservation of Fixed Nitrogen. Eds. J M Lyons, R C Valentine, D A Phillips, D W Rains and R C Huffaker. pp 217–232. Plenum, New York.

Heichel G H, Barnes D K, Vance C P and Henjum K I 1984 N_2 fixation and N and dry matter partitioning during a 4-year alfalfa stand. Crop Sci. 24, 811–815.

Heichel G H, Barnes D K, Vance C P and Hardarson G 1984 Environmental and genotypic effects on dinitrogen fixation of contrasting alfalfa clones. *In* Advances in Nitrogen Fixation Research Eds. C Veeger and W E Newton. pp 595. Nijhoff/Junk Publishers, Boston/Lancaster.

Heichel G H, Hardarson G, Barnes D K and Vance C P 1984 Dinitrogen fixation, herbage yield, and rhizobial preference of selected alfalfa clones. Crop Sci. 24, 1093–1097.

Henson R A and Heichel G H 1984 Dinitrogen fixation of soybean and alfalfa: Comparison of the isotope dilution and difference methods. Field Crops Res. 9, 333–346.

Henzell E F, Martin E A, Ross P J and Haydock K P 1964 Isotopic studies on the uptake of nitrogen by pasture grasses. Aust. J. Agric. Res. 15, 876–884.

Henzell E F, Martin E A, Ross P J and Haydock K P 1968 Isotopic studies on the uptake of nitrogen by pasture plants. Aust. J. Agric. Res. 19, 65–77.

Jensen E 1986 Symbiotic N_2 fixation in pea and field bean estimated by ^{15}N fertilizer dilution in field experiments with barley as a reference crop. Plant and Soil 92, 3–13.

Jensen E 1986 The influence of rate and time of nitrate supply on nitrogen fixation and yield in pea (*Pisum sativum* L.). Fertil. Res. 10, 193–202.

Knowles R 1981 The measurement of nitrogen fixation. *In* Current Perspectives in Nitrogen Fixation. Eds. A H Gibson and W E Newton. pp 327–333. Elsevier Press, Amsterdam.

Kohl D M and Shearer G 1980a Estimation of nitrogen fixation based on differences in the natural abundance of nitrogen-15 in nodulating and non-nodulating isolines of soybeans. Plant Physiol. 66, 61–65.

Kohl D M and Shearer G 1980b Isotopic fractionation associated with symbiotic nitrogen fixation and uptake of nitrate ion by plants. Plant Physiol. 66, 51–56.

Kohl D H and Shearer G 1981 The use of soils highly enriched in nitrogen-15 to screen for nitrogen-fixing activity. Plant and Soil 60, 487–489.

La Favre J S and Focht D D 1983 Comparison of N_2 fixation and yields in *Cajanus cajan* between hydrogenase-positive and hydrogenase-negative rhizobia by *in situ* acetylene reduction assays and direct ^{15}N partitioning. Plant Physiol. 72, 971–977.

Labandera C, Danso S K A, Pastorini D, Curbelo S and Martin V 1988 Nitrogen fixation in a white clover-vescue pasture using three methods of nitrogen-15 application and residual nitrogen-15 uptake. Agron. J. 80, 265–268.

Ladd J N 1981 The use of ^{15}N in following organic matter turnover, with specific reference to rotation systems. Plant and Soil 58, 401–411.

Ladd J N, Oades J M and Amato M 1981 Distribution and recovery of nitrogen from legume residues decomposing in soils sown to wheat in the field. Soil Biol. Biochem. 13, 251–256.

Ladd J N, Amato M, Jackson R B and Butler J H A 1983 Utilization by wheat crops of nitrogen from legume residues decomposing in soils in the field. Soil Biol. Biochem. 15, 231–238.

Ledgard S F, Morton R, Freney J R, Bergersen F J and Simpson J R 1985a Assessment of the relative uptake of added and indigenous soil nitrogen by nodulated legumes and reference plants in the ^{15}N dilution measurement of N_2 fixation: Derivative of method. Soil Biol. Biochem. 17, 317–322.

Ledgard S F, Simpson J R, Freney J R and Bergersen 1985b Field evaluation of ^{15}N techniques for estimating nitrogen fixation in legume-grass associations. Aust. J. Agric. Res. 36, 247–258.

Ledgard S F, Simpson J R, Freney J R, Bergersen F J and Morton R 1985c Assessment of the relative uptake of added and indigenous soil nitrogen by nodulated legumes and reference plants in the ^{15}N dilution measurement of N_2-fixation: Glasshouse application of method. Soil Biol. Biochem. 17, 323–328.

Legg J O and Sloger C 1975 A tracer method for determining symbiotic nitrogen fixation in field studies. *In* Proceedings of the Second International Conference on Stable Isotopes. Eds. E R Klein and P D Klein. pp 661–666. Oak Brook, IL.

Legg J O and Stanford G 1967 Utilization of soil and fertilizer N by oats in relation to the available N status of soils. Soil Sci. Soc. Am. Proc. 31, 215–219.

McAuliffe C, Chamblee D S, Uribe-Arango H and Woodhouse Jr W W 1958 Influence of inorganic nitrogen on nitrogen fixation by legumes as revealed by ^{15}N. Agron. J. 50, 334–337.

Mahendrappa M K, Foster N W, Weetman G F and Krause K K 1986 Nutrient cycling and availability in forest soils. Can. J. Soil Sci. 66, 547–572.

Minchin E R and Sommerfield R J 1978 Potential yield improvement in cowpea (*Vigna unguiculata*): The role of nitrogen nutrition. Ann. Appl. Biol. 88, 468–473.

Minchin F R, Summerfield R J and Eaglesham A R J 1978 Plant genotype × *Rhizobium* strain interactions in cowpea (*Vigna unguiculata* (L.) Walp.). Trop. Agric. Trinidad 55, 107–115.

Papastylianou I and Danso S K A 1989 Effect of nitrogen fertilization and cropping system of the reference crop on estimation of N_2 fixation by vetch using ^{15}N methodology. Plant and Soil 114, 227–233.

Patterson T G and LaRue T A 1983 Nitrogen fixation by soybeans: Seasonal and cultivar effects and comparison of estimates. Crop Sci. 23, 488–492.

Phillips D A and Bennet J P 1978 Measuring symbiotic nitrogen fixation in rangeland plots of *Trifolium subterraneum* L. and *Bromus mollis* L. Agron. J. 70, 671–674.

Phillips D A, Jones M B, Center D M and Vaughn C E 1983a Estimating symbiotic nitrogen fixation by *Trifolium subterraneum* (cultivar Woogenellup) during regrowth. Agron. J. 75, 736–741.

Phillips D A, Center D M and Jones M B 1983b Nitrogen turnover and assimilation during regrowth in *Trifolium subterraneum* L. and *Bromus mollis* L. Plant Physiol. 71, 472–476.

Phillips D A, Jones M B and Foster K W 1986 Advantages of the nitrogen-15 dilution techniques for field measurements of symbiotic dinitrogen fixation in legumes. Soil Sci. Soc. Am. J. 677, 11–21.

Poth M, La Favre J S and Focht D D 1986 Quantification by direct ^{15}N dilution of fixed N_2 incorporation into soil by *Cajanus cajanum* (Pigeon pea). Soil Biol. Biochem. 18, 125–127.

Rao K A and Shinde J E 1984 Turnover of nitrogen-15-labelled urea in various rotations of groundnut, sorghum and pigeon pea crops. Plant and Soil 79, 159–167.

Reichardt K, Hardarson G, Zapata F, Kirda C and Danso S K A 1987 Site variability effect on field measurement of symbiotic nitrogen fixation using the ^{15}N isotope dilution method. Soil Biol. Biochem. 19, 405–409.

Rennie R J 1979 Comparison of ^{15}N-aided methods for determining symbiotic dinitrogen fixation. Rev. Ecol. Biol. Sol. 16, 455–463.

Rennie R J 1982 Quantifying dinitrogen (N_2) fixation in soybeans by nitrogen-15 isotope dilution: The question of the non fixing control plant. Can. J. Bot. 60, 856–861.

Rennie R J 1984 Comparison of N balance and ^{15}N isotope dilution to quantify N_2 fixation in field-grown legumes. Agron. J. 76, 785–790.

Rennie R J 1985 Theoretical errors in sequential ^{15}N labelling of soil to estimate N_2 fixation by isotope dilution. Crop Sci. 25, 891.

Rennie R J 1986 Comparison of methods of enriching a soil with nitrogen-15 to estimate dinitrogen fixation by isotope dilution Agron. J. 78, 158–163.

Rennie R J and Kemp G A 1983 N_2 fixation in field beans quantified by ^{15}N isotope dilution. I. Effects of strains of *Rhizobium phaseoli*. Agron. J. 75, 640–644.

Rennie R J and Kemp G A 1984 ^{15}N-determined time course for N_2 fixation in two cultivars of field beans. Agron. J. 76, 146–154.

Rennie R J, Rennie D A and Fried M 1978 Concepts of ^{15}N usage in dinitrogen fixation studies. *In* Isotopes in Biological Dinitrogen Fixation. IAEA Vienna. pp 107–130.

Rennie R J and Dubetz S 1986 Nitrogen-15-determined nitrogen fixation in field-grown chickpea, lentil, fababean and field pea. Agron. J. 78, 654–660.

Rennie R J and Rennie D A 1983 Techniques for quantifying N_2 fixation in association with non-legumes under field and greenhouse conditions. Can. J. Microbiol. 29, 1022–1035.

Rennie R J, Dubetz S, Bole J B and Muendel H 1982 Dinitrogen fixation measured by nitrogen-15 dilution in 2 Canadian soybean (*Glycine max.*) cultivars. Agron. J. 74, 725–730.

Richards J E and Soper R J 1979 Effect of nitrogen fertilizer on yield, protein content and symbiotic nitrogen fixation in fababean (*Vicia faba*). Agron. J. 71, 807–811.

Ronis D H, Sammons D J, Kenworthy W J and Meisinger J J

1985 Heritability of total and fixed N content of the seed in two soybean populations. Crop Sci. 25, 1–4.

Ruschel A P, Salati E and Vose P B 1979 Nitrogen enrichment of soil and plant by *Rhizobium phaseoli – Phaseolus vulgaris* symbiosis. Plant and Soil 51, 425–429.

Ruschel A P, Vose P B, Victoria R L and Salati E 1979 Comparison of isotope techniques and non-nodulating isolines to study the effect of ammonium fertilization on dinitrogen fixation in soybean, *Glycine max*. Plant and Soil 53, 513–525.

Ruschel A P, Vose P M, Matsui E, Victoria R L and Saito S M 1982 Field evaluation of N_2-fixation and N-utilization by *Phaseolus* bean varieties determined by ^{15}N isotope dilution. Plant and Soil 65, 397–407.

Sanginga N, Mulongoy K and Ayanaba A 1985 Effect of inoculation and mineral nutrients on nodulation and growth of *Leucaena leucocephala*. *In* Biological Nitrogen Fixation in Africa. Eds. H. Ssali and S O Keya. pp 419–436. MIRGEN, Nairobi.

Sanginga N, Zapata F, Danso S K A and Bowen G D 1990 Effect of successive cuttings on uptake and partitioning of ^{15}N among plant parts of *Leucaena leucocephala*. Biol. Fert. Soils 9, 37–42.

Steele K W and Littler R A 1987 Field evaluation of some factors effecting nitrogen fixation in pastures by ^{15}N isotope dilution. Aust. J. Agric. Res. 38, 153–161.

Talbott H J, Kenworthy W J and Legg J O 1982 Field comparison of the nitrogen-15 and difference methods of measuring nitrogen fixation. Agron. J. 74, 799–804.

Vallis I, Haydock K P, Ross P J and Henzell E F 1967 Isotopic studies on the uptake of nitrogen by pasture plants. Aust. J. Agric. Res. 18, 865–877.

Vallis I, Haydock K P, Ross P J and Henzell E F 1977 Uptake of soil nitrogen by legumes in mixed swards. Aust. J. Agric. Res. 28, 413–425.

Van Kessel C, Singleton P W and Hoben H J 1985 Enhanced N-transfer from a soybean to maize by *Vesicular arbuscular* mycorrhizal (VAM) fungi. Plant Physiol. 79, 562–563.

Vasilas B L and Ham G E 1984 Nitrogen fixation in soybeans: An evaluation of measurement techniques. Agron. J. 76, 759–764.

Vasilas B L and Ham G E 1985 Intercropping nodulating and non-nodulating soybeans: Effects on seed characteristics and Dinitrogen fixation estimates. Soil Biol. Biochem. 17, 581–582.

Vose P B and Victoria R L 1986 Re-examination of the limitations of nitrogen-15 isotope dilution technique for the field measurement of dinitrogen fixation. Soil Sci. Soc. Am. J. 677, 23–41.

Vose P B, Ruschel A P, Victoria R L, Saito S M T and Matsui E 1982 Nitrogen-15 as a tool in biological nitrogen fixation research. *In* Biol. Nitrogen Fixation Technol. Trop. Agric. Eds. H Graham and S C Harris. pp 575–592.

Wagner G H and Zapata J 1982 Field evaluation of reference crops in the study of nitrogen fixation by legumes using isotope techniques. Agron. J. 74, 607–612.

Walker T W, Adams A F R and Orchiston H D 1956 Fate of labelled ammonium nitrogen when applied to grass and clover grown separately and together. Soil Sci. 91, 339–351.

Weaver R W 1986 Measurement of biological dinitrogen fixation in the field. *In* Field Measurement of Dinitrogen Fixation and denitrification. Eds. R D Hauck and R W Weaver. pp 1–10. SSSA Special Publication No. 18.

West C P and Wedin W F 1985 Dinitrogen fixation in alfalfa-orchard grass pastures. Agron. J. 77, 89–94.

Williams W A, Jones M B and Delwiche C C 1977 Clover N-fixation measurement by total-N difference and ^{15}N A-values in lysimeters. Agron. J. 69, 1023–1024.

Witty J F 1983 Estimation of N_2-fixation in the field using ^{15}N-labelled fertilizer: Some problems and solutions. Soil Biol. Biochem. 15, 631–640.

Witty J F and Ritz K 1984 Slow-release nitrogen-15 fertilizer formulations to measure nitrogen fixation by isotope dilution. Soil Biol. Biochem. 16, 657–661.

Zapata F, Danso S K A, Hardarson G and Fried M 1987a Nitrogen fixation and translocation in field-grown fababean. Agron. J. 79, 505–509.

Zapata F, Danso S K A, Hardarson G and Fried M 1987b Time course of nitrogen fixation in field-grown soybean using nitrogen-15 methodology. Agron. J. 79, 172–176.

Plant and Soil **152**: 53–58, 1993.

PLSO BNF04

Measurements of N_2 fixation by common bean in Central Brazil as affected by different reference crops

R.A. HENSON
IBTA, Casilla 3299, Cochabamba, Bolivia

Key words: difference method, genotypic variability, ^{15}N isotope dilution, *Phaseolus vulgaris* L.

Abstract

An experiment was conducted at EMBRAPA/CNPAF, Goiânia, Goias, Brazil, on a typic haplustox soil to evaluate growth and N_2 fixation-related parameters of *Phaseolus vulgaris* L. Bean lines, which had been selected for N_2 fixation at CNPAF, including production cultivars, germplasm bank entries, and parents and progenies of a cross made to improve this characteristic. Wheat (*Triticum aestivum* L.) and dwarf sorghum (*Sorghum bicolor* (L.) Moench) were evaluated as non-N_2-fixing reference crops for difference method (DM) and ^{15}N isotope dilution technique (IDT) estimates of N_2 fixation. IDT estimates ranges from 4 to 18 kg N_2 fixed ha^{-1}. High variability associated with low levels of N_2 fixation precluded definitive identification of the best N_2 fixing bean lines. Due to differences in growth cycle and in patterns and amounts of soil N uptake during the season, neither of the reference crops tested appears to be an adequate control for either DM or IDT estimates of N_2 fixation. However, ranking of lines for effectiveness in N_2 fixation could be performed without the use of any reference crops.

Introduction

A non-N_2-fixing reference crop is required for isotope dilution technique (IDT) estimates of N_2 fixation (Rennie et al., 1978). A basic assumption of the IDT is that the reference crop accurately reflects the $^{15}N:^{14}N$ ratio of N absorbed from the soil solution by the N_2-fixing crop.

In most field experiments employing the IDT, the upper portion of the soil profile is amended with isotope-enriched N shortly before or after planting (Hardarson et al., 1984; Henson and Heichel, 1984). The $^{15}N:^{14}N$ ratio of plant-available N subsequently falls quickly as the applied isotope is incorporated into the soil organic matter pool (Caballero et al., 1986). Thus, the pattern of soil N uptake during the growing season by N_2-fixing and reference crops is critical. If one crop absorbs N earlier or later in the season relative to the other, IDT estimates of fixed N_2 will be confounded by the changing $^{15}N:^{14}N$ ratio of soil-derived N. A further consideration is the rooting pattern of N_2-fixing and reference crops. If both crops do not absorb the same proportions of total plant N from the isotope-enriched zone of the soil profile, IDT estimates will again be confounded. Both of the above parameters, i.e. N uptake and rooting patterns, will particularly influence N_2 fixation measurements when the effectiveness is low (Hardarson et al., 1988).

In soybean (*Glycine max* (L.) Merr.), the discovery of a non-nodulating type (Williams and Lynch, 1954) and subsequent development of non-nodulating isolines of commercial varieties have facilitated the quantification of fixed N_2. Although shown to be a less than ideal reference crop under some growing conditions (Deibert et al., 1979; Rennie, 1982), the use of non-nodulating isolines has contributed to reduce differences between soybeans and reference crops in N absorption and rooting patterns (Bhangoo and Albritton, 1976; Henson and Heichel, 1984).

The quantification of N_2 fixation by common bean (*Phaseolus vulgaris* L.) has been limited by the presence of indigenous, sometimes ineffective *Rhizobium leguminosarum* biovar *phaseoli* in most bean production soils, the lack of non-nodulating isolines of test cultivars, and problems resulting from differences in N uptake rates and rooting patterns between beans and reference crops. A non-nodulating bean genotype has been identified (Davis et al., 1988), but until this characteristic has been incorporated into locally-adapted germplasm, researchers continue to depend on reference crops from widely divergent crops, mostly grasses. Normally, samples are taken only at bean physiological maturity and estimates of N_2 fixation are calculated without regard for possible differences between N_2 fixing and reference crops in N uptake and rooting pattern.

The objectives of this research were to: evaluate non-N_2-fixing reference crops for use in DM and IDT estimates of N_2 fixation and quantify the mass and proportion of fixed N_2 and soil-derived N in production cultivars and in elite bean genotypes selected for high nodulation, N accumulation and yield.

Materials and methods

A field experiment was planted in 1987 at the EMBRAPA National Center for Rice and Bean Research (CNPAF), Goiânia, Goias, Brazil, to quantify the N_2 fixed by 17 elite bean genotypes selected for N_2 fixation at CNPAF. Entries included: 'Rio Tibagi' and 'Carioca' (poor and good N_2-fixing production cultivars, respectively); Negro Argel, Preto Cariri, Honduras 35, and Venezuela 350 PS (selections from the CNPAF Active Germplasm Bank); WBR 22-3, WBR 22-8, WBR 22-14, WBR 22-24, WBR 22-27, WBR 22-34, WBR 22-50, WBR 22-52, and WBR 22-55 (sister lines bred specifically for increased N_2 fixation capacity) (Attewell and Bliss, 1985; Bliss et al., 1989; McFerson et al., 1982); and ICA Pijao and Puebla 152 (good agronomic and N_2-fixing, respectively, parental materials of these sister lines). The variety Carioca is of the 'carioca' seed type, all other entries are of the black-seeded type. All lines are of plant type II except Puebla 152, which has a type III growth habit (CIAT, 1987). Concurrently, two grass species were evaluated as DM and IDT non-N_2-fixing reference crops: dwarf sorghum (*Sorghum bicolor* (L.) Moench) line BR 005 (planted in both solid-seeded rows and in hills at the same intrarow plant spacing as the beans) and wheat (*Triticum aestivum* L.) line BR 10 (planted in solid-seeded rows).

Prior to planting beans, the experimental area was cropped to one cycle of maize (*Zea mays* L.). In mid-July, the experiment was installed in the CNPAF bean research area (red-yellow latosol, typic haplustox) in a randomized complete block design with six replicates. A 1.5 m section of each one-row × 0.5 m plot was labelled with a surface application of $(^{15}NH_4)_2SO_4$ (10.0% ^{15}N atom excess) at the rate of 0.100 g $^{15}N\ m^{-2}$ (10 kg N ha^{-1}) dissolved in four liters water m^{-2}. Natural abundance $(NH_4)_2SO_4$ was applied to a 13.5 m extension of the isotope-treated area in each row. Overhead irrigation was then applied to move the N down into the rooting zone. The entire experimental area was fertilized with 80 kg P_2O_5 ha^{-1}, 60 kg K_2O ha^{-1}, and 0.5 kg MoO_4 ha^{-1} applied as granular fertilizer in the row one day before planting. Beans were seed-innoculated with a peat-based mixture of four strains of *R. leguminosarum* biovar *phaseoli* selected at CNPAF (CNPAF 150, CNPAF 160, CNPAF 170 and CNPAF 189). Ten days after emergence (DAE), beans were thinned to 20 plants m^{-2}.

A five-plant sample was collected from the natural abundance areas of each plot at two vegetative stages (V 2–3 and V 5–6) (Lebaron, 1974) and the early reproductive stage (R 1–4) to determine shoot, root, and nodule dry matter accumulation, and shoot total N content. At physiological maturity (66 DAE; R 7), 1.2 m of each row (0.6 m^2) was collected from the isotope-enriched area of each plot to determine shoot dry matter content, total N and ^{15}N enrichment. On that date, bean shoots were divided into pods and stems + leaves for separate analyses. Weighted data from these subsamples were combined for whole shoot values. Subsamples of the material collected from ^{15}N-labelled areas were analyzed for ^{15}N:^{14}N ratio at the IAEA Seibersdorf Laboratory. Total N Kjeldahl analyses were conducted at CNPAF. DM estimates of fixed N_2 were based upon values for each grass treatment and for the bean line with the lowest shoot N mass (Rio Tibagi at 19 DAE, WBR 22–34 at 33 DAE, and ICA Pijao at 51 and 66 DAE).

Results and discussion

Nodulation in this experiment was relatively low, and by the mid-reproductive phase (51 DAE) most nodules had senesced (Table 1). Negro Argel and WBR 22–24 produced the best early-season nodulation, but no statistically significant ($p < 0.05$) differences in nodule dry weight were detected after 19 DAE.

At 66 DAE, WBR 22–34, Venezuela 350 PS, and Negro Argel contained the highest whole shoot dry matter (Table 2). Honduras 35, Carioca, and Negro Argel were the most efficient in converting whole shoot dry matter into pod dry matter, while Preto Cariri and Puebla 152 were the least efficient. Wheat shoot dry matter through 66 DAE was similar to that of the beans, while sorghum contained a significantly higher shoot dry matter than the beans on this date.

The weight of total shoot N (Table 3) generally reflected shoot dry matter content, but Preto Cariri and Puebla 152 also displayed a relatively high shoot N content at 66 DAE. Honduras 35 accumulated the highest amount of N during early growth, while Puebla 152 was consistently high throughout the season. WBR 22–34 and Preto Cariri accumulated high amounts of N during the reproductive period. ICA Pijao contained a relatively low amount of shoot N throughout the season. Among the grasses, wheat accumulated N faster than the beans during the vegetative phase, but was surpassed by almost all bean entries during reproductive growth. Solid-seeded sorghum accumulated amounts of N statistically similar to the beans during vegetative growth, while hilled sorghum accumulated N very slowly during the early season. The N accumulation by both sorghum densities far surpassed that of beans by 66 DAE.

Table 1. Nodule dry weight (mg/plant) in 17 cultivars of common bean at three harvest days. Values are means of six replicates

Genotype	DAE		
	19	33	51
Rio Tibagi	15 C[a]	24 A	3 A
Carioca	37 ABC	39 A	3 A
Negro Argel	48 A	40 A	15 A
Preto Cariri	25 ABC	45 A	9 A
Puebla 152	30 ABC	51 A	17 A
ICA Pijao	24 BC	31 A	9 A
Honduras 35	35 ABC	36 A	5 A
WBR 22–3	33 ABC	54 A	1 A
WBR 22–8	31 ABC	38 A	6 A
WBR 22–14	22 BC	38 A	11 A
WBR 22–24	39 AB	49 A	12 A
WBR 22–34	20 BC	37 A	9 A
WBR 22–50	27 ABC	35 A	3 A
WBR 22–52	28 ABC	40 A	8 A
WBR 22–55	24 BC	43 A	6 A
Ven. 350 PS	31 ABC	37 A	5 A
WBR 22–27	25 ABC	29 A	3 A
$\bar{X}$	29	39	7

[a] Means within a column followed by the same letter do not differ significantly ($p < 0.05$) by Tukey's HSD.

Table 2. Dry matter content (g/plant) of whole shoots and pods of 17 cultivars of common beans and three reference crops by 66 DAE. Values are means of six replicates

Genotype	Whole shoot		Pods
Rio Tibagi	11.18 C[a]	11.18 ABC	5.50 CDEF
Carioca	13.00 C	13.00 ABC	8.27 AB
Negro Argel	13.53 C	13.53 AB	8.30 AB
Preto Cariri	13.36 C	13.36 ABC	3.73 F
Puebla 152	13.21 C	13.21 ABC	5.19 DEF
ICA Pijao	9.39 C	9.39 C	4.87 EF
Honduras 35	12.36 C	12.36 ABC	8.84 A
WBR 22–3	11.76 C	11.76 ABC	6.32 BCDE
WBR 22–8	11.41 C	11.41 ABC	5.54 CDEF
WBR 22–14	12.32 C	12.32 ABC	6.34 BCDE
WBR 22–24	11.17 C	11.17 ABC	6.12 BCDE
WBR 22–34	14.41 C	14.41 A	7.27 ABCD
WBR 22–50	9.69 C	9.69 BC	4.98 EF
WBR 22-52	10.43 C	10.43 ABC	5.07 DEF
WBR–22–55	11.58 C	11.58 ABC	6.23 BCDE
Ven. 350 PS	13.88 C	13.88 A	7.47 ABC
WBR 22–27	11.37 C	11.37 ABC	6.03 CDE
Wheat [b]	14.36 C	–	–
Solid sorghum[b]	43.77 A	–	–
Hilled sorghum	32.19 B	–	–
$\bar{X}$	14.71	12.00	6.24

[a] Means within a column followed by the same letter do not differ significantly ($p < 0.05$) by Tukey's HSD.
[b] Values represent the area occupied by one bean plant ($0.05\ m^2$).

Table 3. Shoot N content (mg/plant) in 17 common bean cultivars and three reference crops at four harvest days. Values are means of six replicates

Genotype	DAE			
	19	33	51	66
Rio Tibagi	16 CD[a]	81 B	221 A	200 B
Carioca	25 BC	88 B	248 A	234 B
Negro Argel	25 BC	94 B	291 A	238 B
Preto Cariri	24 BC	86 B	314 A	261 B
Puebla 152	26 BC	106 AB	302 A	249 B
ICA Pijao	21 BC	76 B	201 A	166 B
Honduras 35	35 B	94 B	311 A	228 B
WBR 22–3	25 BC	101 B	226 A	207 B
WBR 22–8	28 BC	83 B	206 A	203 B
WBR 22–14	19 CD	76 B	214 A	233 B
WBR 22–24	24 BC	88 B	230 A	192 B
WBR 22–34	21 BC	74 B	263 A	259 B
WBR 22–50	21 BC	80 B	240 A	178 B
WBR 22–52	24 BC	84 B	202 A	195 B
WBR 22–55	23 BC	88 B	234 A	210 B
Ven. 350 PS	18 CD	84 B	229 A	257 B
WBR 22–27	21 BC	93 B	215 A	212 B
Wheat [2]	52 A	155 A	226 A	179 B
Solid sorghum[b]	30 BC	112 AB	302 A	573 A
Hilled sorghum	5 D	62 B	287 A	514 A
$\bar{X}$	24	90	248	249

[a] Means within a column followed by the same letter do not differ significantly ($p < 0.05$) by Tukey's HSD.
[b] Values represent the area occupied by one bean plant (0.05 m^2).

On this date (bean physiological maturity), the sorghum was in the flowering stage of growth.

DM estimates based on wheat at 66 DAE showed low values of N_2 fixation: 25% of total shoot N and 12 kg ha^{-1} across bean genotypes (Table 4). Solid-seeded sorghum gave negative values at 66 DAE, and hilled sorghum gave positive, but very low values. Since ICA Pijao did form nodules, it is assumed that it also fixed N_2 and that DM estimates based on this cultivar are underestimates. However, these calculations serve to establish a minimum value of N_2 fixed by the other entries. By this method, the beans fixed an average of at least 11 kg N_2 ha^{-1} (24% of total shoot N).

The ^{15}N atom excess at 66 DAE was higher in grasses than in beans (Table 5) indicating significant ($p < 0.05$) N_2 fixation. No statistically significant differences in ^{15}N concentration were detected among beans.

IDT estimates of fixed N_2 ranged from 4 to 18 kg ha^{-1} and from 12 to 36% of total shoot N derived from fixation (Table 6). The N_2 fixation was probably so low due to warm temperatures during the period of reproductive stage which is likely to have induced nodule senescence and decreased nitrogen fixation. Under conditions of

Table 4. N difference method estimates of N_2 fixation by 17 common bean cultivars at 66 DAE using several reference crops. Values are means of six replicates

Genotype	Wheat	Sorghum		ICA Pijao	Wheat	Sorghum		ICA Pijao
		Solid	Hills			Solid	Hills	
		(%)			(kg ha^{-1})			
Rio Tibagi	23 AB*	0 A	8 A	20 ABC	10 AB	0 A	4 A	9 AB
Carioca	29 AB	0 A	6 A	29 AB	15 AB	0 A	2 A	14 AB
Negro Argel	31 A	0 A	9 A	29 AB	15 AB	0 A	5 A	14 AB
Preto Cariri	31 A	0 A	5 A	36 A	17 AB	0 A	2 A	19 A
Puebla 152	32 A	0 A	8 A	33 AB	17 AB	0 A	4 A	17 A
ICA Pijao	12 B	0 A	0 A	0 C	4 B	0 A	0 A	0 B
Honduras 35	26 AB	0 A	7 A	28 AB	12 AB	0 A	3 A	13 AB
WBR 22–3	24 AB	0 A	3 A	22 AB	11 AB	0 A	1 A	9 AB
WBR 22–8	23 AB	0 A	2 A	20 ABC	12 AB	0 A	0 A	9 AB
WBR 22–14	28 AB	0 A	7 A	29 AB	13 AB	0 A	3 A	13 AB
WBR 22–24	20 AB	0 A	4 A	16 ABC	8 AB	0 A	1 A	6 AB
WBR 22–34	35 A	0 A	8 A	36 A	19 A	0 A	4 A	19 A
WBR 22–50	18 AB	0 A	4 A	14 BC	7 AB	0 A	1 A	5 AB
WBR 22–52	21 AB	0 A	2 A	17 ABC	9 AB	0 A	1 A	7 AB
WBR 22–55	23 AB	0 A	5 A	22 AB	11 AB	0 A	2 A	9 AB
Ven. 350 PS	29 AB	0 A	6 A	31 AB	18 AB	0 A	2 A	18 A
WBR 22–27	25 AB	0 A	7 A	21 ABC	12 AB	0 A	3 A	11 AB
$\bar{X}$	25	0	5	24	12	0	2	11

* Means within a column followed by the same letter do not differ significantly ($p < 0.05$) by Tukey's HSD.

Table 5. ^{15}N concentration in whole shoots of 17 common bean cultivars and three reference crops at 66 DAE. Values are means of six replicates

Genotype	% ^{15}N atom excess
Rio Tibagi	0.276 C*
Carioca	0.268 C
Negro Argel	0.254 C
Preto Cariri	0.291 BC
Puebla 152	0.263 C
ICA Pijao	0.275 C
Honduras 35	0.253 C
WBR 22–3	0.269 C
WBR 22–8	0.269 C
WBR 22–14	0.274 C
WBR 22–24	0.275 C
WBR 22–34	0.261 C
WBR 22–50	0.283 BC
WBR 22–52	0.312 ABC
WBR 22–55	0.266 C
Ven. 350 PS	0.270 C
WBR 22–27	0.330 ABC
Wheat	0.402 A
Solid sorghum	0.350 ABC
Hilled sorghum	0.380 AB
$\bar{X}$	0.291

* Means within a column followed by the same letter do not differ significantly ($p < 0.05$) by Tukey's HSD.

low N_2 fixation, variability of estimates increases (Hardarson et al., 1988) and essentially no statistically significant variation among bean genotypes was found. Although there are all indications that N_2 fixation was very low in this experiment, the apparent lack of an appropriate control crop makes the accuracy of these quantification estimates questionable. Although the planting to harvest cycles of wheat and beans were similar, shoot total N content of wheat peaked during the early reproductive phase and then declined (Table 4), while beans normally continue to accumulate N up to growth stage R 7. In this experiment, bean shoot N content also declined from 51 DAE to 66 DAE, due to leaf loss from slight insect damage and probable high respiration due to warm temperatures during this period. The growth cycle of dwarf sorghum was considerably longer than the beans and the pattern and amount of N uptake over the season were also quite different. Thus, the pattern of N accumulation by all grass treatments in this experiment was quite different from that normally observed in beans and none of the grass treatments appears to be an adequate control for

Table 6. Isotope dilution technique estimates of N_2 fixed by 17 common bean cultivars at 66 DAE using several non-fixing reference crops. Values are means of six replicates

Genotype	Control crop					
	Wheat	Sorghum		Wheat	Sorghum	
		Solid	Hills		Solid	Hills
	(%)			(kg ha^{-1})		
Rio Tibagi	30 A[a]	22 A	25 A	12 A	8 A	10 AB
Carioca	33 A	24 A	29 A	16 A	11 A	13 AB
Negro Argel	36 A	24 A	32 A	17 A	12 A	15 A
Preto Cariri	27 A	15 A	23 A	15 A	8 A	12 AB
Puebla 152	34 A	23 A	30 A	17 A	11 A	15 AB
ICA Pijao	31 A	22 A	26 A	11 A	8 A	9 AB
Honduras 35	36 A	25 A	32 A	17 A	12 A	15 AB
WBR 22–3	33 A	20 A	28 A	14 A	9 A	12 AB
WBR 22–8	33 A	21 A	28 A	14 A	9 A	11 AB
WBR 22–14	31 A	22 A	27 A	14 A	10 A	12 AB
WBR 22–24	31 A	22 A	27 A	12 A	8 A	10 AB
WBR 22–34	34 A	22 A	29 A	18 A	11 A	15 A
WBR 22–50	29 A	19 A	24 A	10 A	6 A	8 AB
WBR 22–52	21 A	14 A	16 A	8 A	6 A	6 AB
WBR 22–55	33 A	21 A	29 A	14 A	9 A	12 AB
Ven. 350 PS	32 A	20 A	28 A	17 A	9 A	13 AB
WBR 22–27	20 A	12 A	15 A	8 A	4 A	5 B
$\bar{X}$	31	20	26	14	9	11

[a] Means within a column followed by the same letter do not differ significantly ($p < 0.05$) by Tukey's HSD.

either DM or IDT estimates of N_2 fixation. However, % ^{15}N atom excess can be used to rank genotypes in terms of effectiveness. In this case the genotype with the lowest percentage of ^{15}N a.e. would rank as the most efficient in nitrogen fixation, i.e., Honduras 35 followed by Negro Argel, WBR 22–34, Puebla 152 and WBR 22–55 (Table 5). This type of ranking can be done without the use of any reference crop.

Acknowledgements

The author wishes to thank F A Bliss, P A A Pereira and R S Araujo for scientific collaboration and A F Borges, H D de Rezende, J S de Rezende, T M Damasceno, D F Bastos, P T Lobo, Elsa, and A F dos Santos for capable technical assistance. This work was sponsored in part by the Empresa Brasileira de Pesquisa Agropecuaria (EMBRAPA); the Bean/Cowpea CRSP funded by USAID Grant No. DAN-1310-G-SS-6008-00; and by the International Atomic Energy Agency contract No. 4250/R1/RB.

References

Attewell J and Bliss F A 1985 Host plant characteristics of common bean lines selected using indirect measures of N_2 fixation. *In* Nitrogen Fixation Research Progress. Eds. H J Evans, P J Bottomly and W E Newton. pp 3–9. Martinus Nijhoff, Dordrecht.

Bhangoo M S and Albritton D J 1976 Nodulating and non-nodulating Lee soybean isolines response to applied nitrogen. Agron. J. 68, 642–645.

Bliss F A, Pereira P A A, Araujo R S, Henson R A, Kmiecik K A, McFerson J R, Teixeira M G and da Silva C C 1989 Registration of five high nitrogen-fixing common bean germplasm lines. Crop Sci. 29, 240–241.

Caballero S S U, Libardi P L, Reichardt K, Moraes O and Victoria R L 1986 Lixiviaçao do nitrogenio proveniente do solo e do fertilizante $(^{15}NH_4)_2SO_4$ durante o ciclo de uma cultura de feijao. Pesquisa Agropecuaria Brasileira 21, 25–31.

CIAT (Centro Internacional de Agricultura Tropical) 1987 Sistema estandar para la evaluacion de germoplasma de frijol. Comps. A van Schoonhoven and Pastor-Corrales M A, Cali, Colombia. 56 p.

Davis J H C, Giller K E, Kipe-Nolt J and Awah M 1988 Non-nodulating mutants in common bean. Crop Sci. 28, 859–860.

Deibert E J, Bijeriego M and Olson R A 1979 Utilization of ^{15}N fertilizer by nodulating and non-nodulating soybean isolines. Agron. J. 71, 717–723.

Hardarson G, Zapata F and Danso S K A 1984 Field evaluation of symbiotic nitrogen fixation by rhizobial strains using ^{15}N methodology. Plant and Soil 82, 369–375.

Hardarson G, Danso S K A and Zapata F 1988 Dinitrogen fixation measurements in alfalfa-ryegrass swards using nitrogen-15 and influence of the reference crop. Crop Sci. 28, 101–105.

Henson R A and Heichel G H 1984 Dinitrogen fixation of soybean and alfalfa: Comparison of the isotope dilution and difference methods. Field Crops Res. 9, 333–346.

Lebaron M J 1974 Developmental stages of the common bean plant. University of Idaho College of Agriculture, Current Information Series No. 288. 2 p.

Legg J O and Sloger C 1975 A tracer method for determining symbiotic nitrogen fixation in field studies. *In* Proc. 2nd Int. Conf. on Stable Isotopes, Oak Brook, IL. Eds. E R Klein and P D Klein. pp 661-666. U.S. Energy Research and Development Administration, Washington DC.

McFerson J, Bliss F A and Rosas J C 1982 Selection for enhanced nitrogen fixation in common beans, *Phaseolus vulgaris*. *In* Biological Nitrogen Fixation Technology for Tropical Agriculture. Eds. P H Graham and S C Harris. pp 39–44. CIAT, Cali, Colombia.

Rennie R J 1982 Quantifying dinitrogen (N_2) fixation in soybeans by ^{15}N isotope dilution: The question of the non-fixing control plant. Can. J. Bot. 60, 856–861.

Rennie R J, Rennie D A and Fried M 1978 Concepts of ^{15}N usage in dinitrogen fixation studies. *In* Isotopes in Biological Dinitrogen Fixation. pp 107–133. International Atomic Energy Agency, Vienna.

Weber C R 1966 Nodulating and non-nodulating soybean isolines I. Agronomic and chemical attributes. Agron. J. 58, 43–46.

Williams L F and Lynch D L 1954 Inheritance of a non-nodulating character in soybeans. Agron. J. 46, 28–29.

Plant and Soil **152**: 59–70, 1993.

PLSO BNF05

Genotypic variation in biological nitrogen fixation by common bean

G. HARDARSON[1], F.A. BLISS[2], M.R. CIGALES-RIVERO[3], R.A. HENSON[4], J.A. KIPE-NOLT[5], L. LONGERI[6], A. MANRIQUE[7], J.J. PEÑA-CABRIALES[8], P.A.A. PEREIRA[4], C.A. SANABRIA[9] and S.M. TSAI[10]

[1]*FAO/IAEA Laboratory, A-2444 Seibersdorf, Austria,* [2]*Department of Pomology, University of California-Davis, CA 95616-8683, USA,* [3]*University Center for Research and Agricultural Development (CUIDA), Apartado Postal 199 28000, Colima, Col., Mexico,* [4]*CNPAF/EMBRAPA, Department of Microbiology, Caixa Postal 179, 74000 Goiania-Goias, Brazil,* [5]*CIAT, Apartado Aéreo 6713, Cali, Colombia,* [6]*Universidad de Concepción, Casilla 537, Chillán, Chile,* [7]*Universidad Nacional Agraria-La Molina, Apartado 456, Lima, Perú,* [8]*CINVESTAV, Apartado Postal 629, 36500 Irapuato, Gto., Mexico,* [9]*DGEN, Guatemala City, Guatemala,* [10]*CENA, Av.Centenario 303, Bairro Sao Dimas, 13416-000 Piracicaba, Sao Paulo, Brazil*

Key words: common bean, N_2, ^{15}N isotope dilution, nitrogen fixation, *Phaseolus vulgaris* L., *Rhizobium*

Abstract

Field experiments were performed in Austria, Brazil, Chile, Colombia, Guatemala, Mexico and Peru as part of an FAO/IAEA Co-ordinated Research Programme to investigate the nitrogen fixing potential of cultivars and breeding lines of common bean (*Phaseolus vulgaris* L.). Each experiment included approximately 20 bean genotypes which were compared using the ^{15}N isotope dilution method. Great differences in nitrogen fixation were observed between and within experiments, with average values of 35% N derived from atmosphere (% Ndfa) and highest values of 70% Ndfa being observed. These values which were larger than had been reported previously for common bean, were observed only when environmental factors were favorable. Therefore, common bean lines are available, which can support high biological nitrogen fixation. These can be used either directly as cultivars for production or in breeding programmes to enhance nitrogen fixation in other cultivars.

Introduction

The effectiveness of various legume species and their micro-symbionts has been listed in several publications (FAO, 1984; Hardarson et al., 1987). For example, fababean (*Vicia faba*), lupin (*Lupinus* spp.) and pigeon peas (*Cajanus cajan*) have been found to be very efficient (>70% N derived from atmosphere, % Ndfa); soybean (*Glycine max*), ground nut (*Arachis hypogea*) and cowpea (*Vigna unguiculata*) average (40–60% Ndfa); and common bean (*Phaseolus vulgaris*) and pea (*Pisum sativum*) on the average rather poor (approximately 30% Ndfa) in fixing atmospheric nitrogen.

An FAO/IAEA Consultants Meeting held 26–30 September 1983, recommended the initiation of a research programme to enhance biological nitrogen fixation by common bean (Hardarson and Lie, 1984). After this meeting an FAO/IAEA Co-ordinated Research Programme on 'The Use of Isotopes in Studies to Improve Yield and Nitrogen Fixation of Common Bean (*Phaseolus vulgaris* L.) in Latin America' was initiated in 1986 and was completed in 1991. The study, which is being reported here, was performed as part of the above programme. Its objectives were to investigate the N_2 fixation potential of various cultivars and breeding lines of common bean and to identify lines, which could be used as parents in breeding programmes to enhance N_2 fixation in this species.

Materials and methods

All participants conducted similar types of experiments to quantify biological nitrogen fixation in variable numbers of common bean cultivars and breeding lines using the ^{15}N isotope dilution method (Fried and Middelboe, 1977; Hardarson, 1990; McAuliffe et al., 1958). The lines tested as well as non-fixing reference crops are listed in Tables 1–8. The seeds were inoculated with the best available strains of *Rhizobium leguminosarum* biovar. *phaseoli* at each location. Every experiment had six replicates of each common bean entry and the non-fixing standard crop grown in 1.5 m rows in a randomized complete block design. Each row had an inter-row plant spacing of 10 cm and 40 cm between rows. 10 kg N/ha of approximately 10% ^{15}N atom excess ammonium sulphate was applied to the experimental area at the time of planting. Plants were cared for by providing the usual practices for production in each region.

Only cultivars of similar growth periods were included in each experiment. The evaluations of nitrogen fixation were made when the cultivars having the shortest growth cycle reached physiological maturity. An exception was made in Chile where the cultivars were grouped according to types. At physiological maturity of the earliest cultivar all entries were harvested. Plant samples were weighed, chopped into 1–2 cm pieces, sub-sampled and dried at 70°C. The dried samples were ground to pass through a 0.5 mm sieve. Total N determinations were done by the Kjeldahl method (Eastin, 1978), and the N isotope ratio analyses by mass spectrometer (Fiedler and Proksch, 1975) at the IAEA's Seibersdorf Laboratory in Austria. The calculation of N_2 fixation was performed using the isotope dilution method (Fried and Middelboe, 1977; McAuliffe et al., 1958). The measurements of % Ndfa and amount of nitrogen fixed was done using a suitable non-fixing reference crop.

Table 1. Common bean cultivars tested by Hardarson (Austria)

No.	Cultivars
1[a]	Riz 30
2	Dor 41
3	Riz 29
4	Riz 22
5	A 237
6	Bat 332
7	Riz 68
8	Rojo 70
9	Riz 53
10	Catu
11	Bat 1645
12	Puebla 152
13	Riz 13
14	Riz 27
15	Riz 36
16	Riz 34
17	Riz 10
18	Wte 3
19	Riz 32
20	Bat 76
21	Bat 477
22	Bat 22
23	Riz 44
24	Tostada
25	Nmutikili
26	Rubona
27	Ikinimba
28	Borlotto
29	Extender
Ref.	Non-nod soybean

[a] 1–23 from CIAT, Colombia; 24–27 from Rwanda; 28–29 from Austria.

Results

Austria

Twenty-nine bean lines and non-nodulating soybean used as the non-fixing reference crop were grown at Seibersdorf, Austria (Table 1). The % Ndfa ranged from 27 to 67% and total fixed N from 25 to 165 kg ha^{-1}. The best cultivars for % Ndfa were Riz 44 and BAT 322 at approx. 65% (Fig. 1A), while Rubona and Rojo 70 fixed about 165 kg N ha^{-1} (Fig. 1B). There were no strong correlations between the two measures of N_2 fixation. Three lines that fixed large amounts of total Ndfa ha^{-1} were from Rwanda (Nos. 26, 24 and 27).

Brazil

Seventeen bean lines and sorghum used as the reference crop were grown at Goiania, Brazil (Table 2). The values for % Ndfa were low, ranging from 12 to 25%, with the best lines being Honduras 35 and Carioca (Fig. 2A). The highest amounts of fixed N were 11–12 kg ha^{-1} by

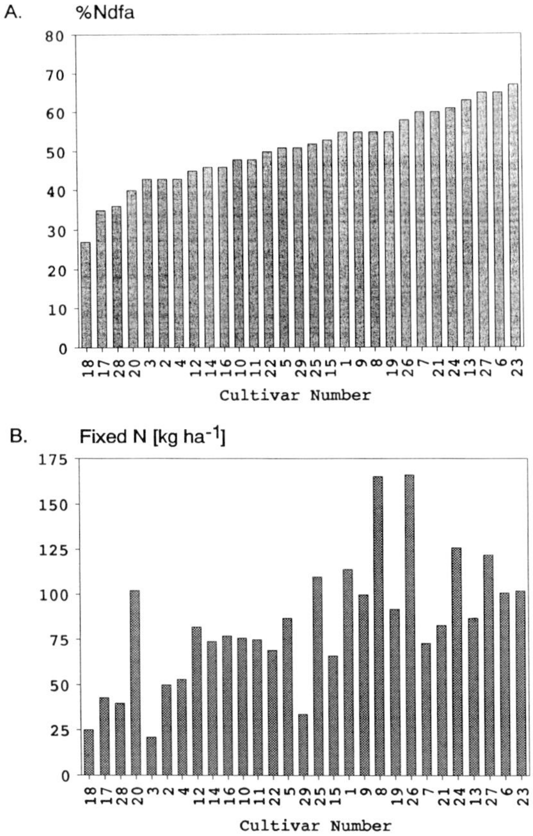

Fig. 1. (**A**) %N derived from atmosphere (%Ndfa) of common bean cultivars tested under field conditions in Austria by G. Hardarson and (**B**) the amount of N_2 fixed by the same cultivars. The cultivars are listed in Table 1.

Table 2. Common bean cultivars tested by Henson and Tsai in Brazil, 1987

Henson		Tsai	
No.	Cultivars	No.	Cultivars
1	Rio Tibagi	1	P2PVBZ 1637
2	Carioca	2	Mexico 309
3	Negro Argel	3	Carioca 80
4	Preto Cariri	4	Negro Argel
5	Puebla 152	5	A230
6	ICA Pijao	6	WBR22-34
7	Honduras 35	7	Rio Tibagi
8	WBR 22-3	Ref.	
9	WBR 22-8		
10	WBR 22-14		
11	WBR 22-24		
12	WBR 22-34		
13	WBR 22-50		
14	WBR 22-52		
15	WBR 22-55		
16	Venezuela 350 PS		
17	WBR 22-27		
Ref.	Sorghum		

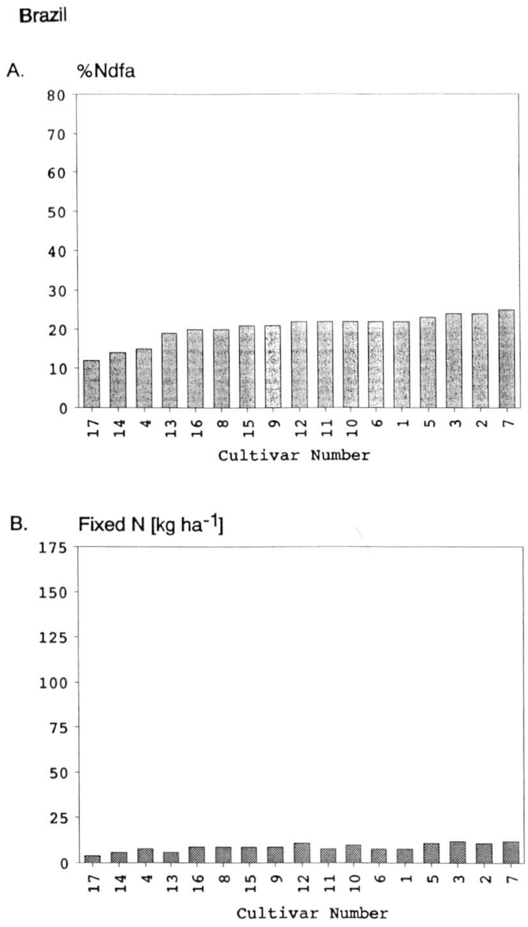

Fig. 2. (**A**) %N derived from atmosphere (%Ndfa) of common bean cultivars tested under field conditions in Brazil by R A Henson and P A A Pereira and (**B**) the amount of N_2 fixed by the same cultivars. The cultivars are listed in Table 2.

Carioca (No. 2), Negro Argel (No. 3), and Puebla 152 (No. 5).

Among seven lines grown at Piracicaba, Brazil, Negro Argel (53%) and Mexico 309 (Table 2, Fig. 3) (51%) produced the highest values for % Ndfa, and for total fixed N (45 and 32 kg ha^{-1}, respectively). There was a positive relationship between % Ndfa and total Ndfa for the lines grown at Piracicaba.

Chile

The experiment conducted in Chile in 1987 included seven lines in each of the type I, II and III growth habit classifications and the non-nod soybean used as the non-fixing reference crop

Table 3. Common bean cultivars tested by Longeri (Chile)

Type	1987 No.	1987 Cultivars	1988 No.	1988 Cultivars
I	1	Blanco INIA	1	Pinto 114
	3	Fleetwood	2	Blanco INIA
	5	Frutilla Canete	3	Laker
	6	GN 114	4	Harrofleet
	7	Pinto 114		
	8	Quila 303		
	9	Red Mexican INIA		
	Ref.	Non-nod soybean		
II	11	Black Mexican	5	Orfeo INIA
	12	Don Timoteo	6	G.N. Harris
	14	Negro Argel	7	Pinto Gala
	15	Negro 115	8	Opalo
	16	Orfeo INIA	9	G.N. Valley
	17	Suave INIA		
	18	Tortola INIA		
	Ref.	Non-nod soybean		
III	21	Araucano INIA	10	Araucano INIA
	22	Burros	11	Coscorron Granado INIA
	23	Burroa Argentinos	12	Tortola Corriente
	24	Cuyano INIA	Ref.	Non-nod soybean
	25	Frutilla Corriente		
	29	Sapito		
	30	Suaves		
	Ref.	Non-nod soybean		

(Table 3). Similar ranges from 35 to 55% Ndfa were seen in each of the three plant type groups, with the highest values being in Red Mexican INIA (Type I) and Don Timoteo (Type II) (Fig. 4A). The highest level of Ndfa was 62 kg N ha^{-1}, with at least one line in each of the three classes fixing at least 50 kg N ha^{-1} (Fig. 4B). Twelve lines, some the same as tested in 1987, were grown in 1988, with % Ndfa ranging from 27 to 60% and total Ndfa from 25 to 115 kg ha^{-1} (Fig.

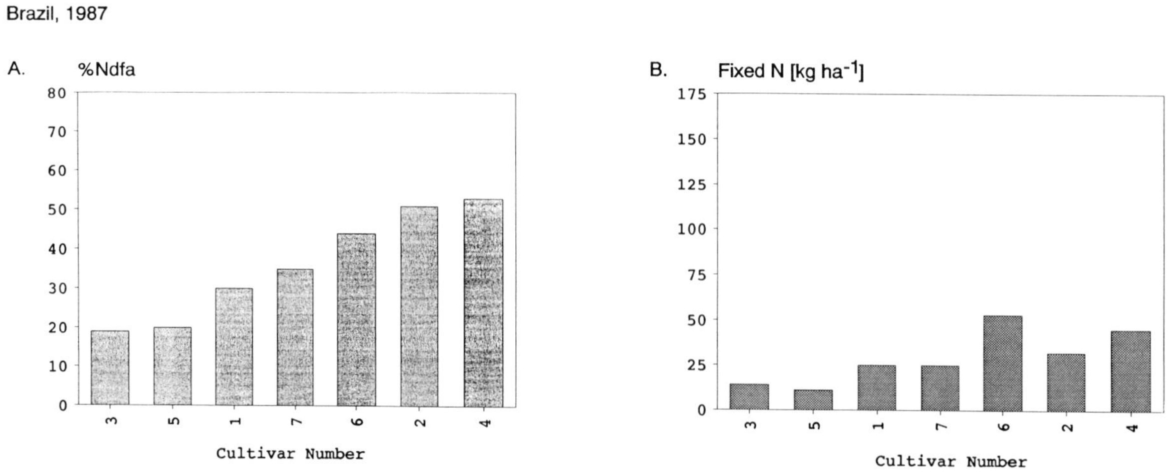

Fig. 3. (**A**) %N derived from atmosphere (%Ndfa) of common bean cultivars tested under field conditions in Brazil by S M Tsai and (**B**) the amount of N_2 fixed by the same cultivars. The cultivars are listed in Table 2.

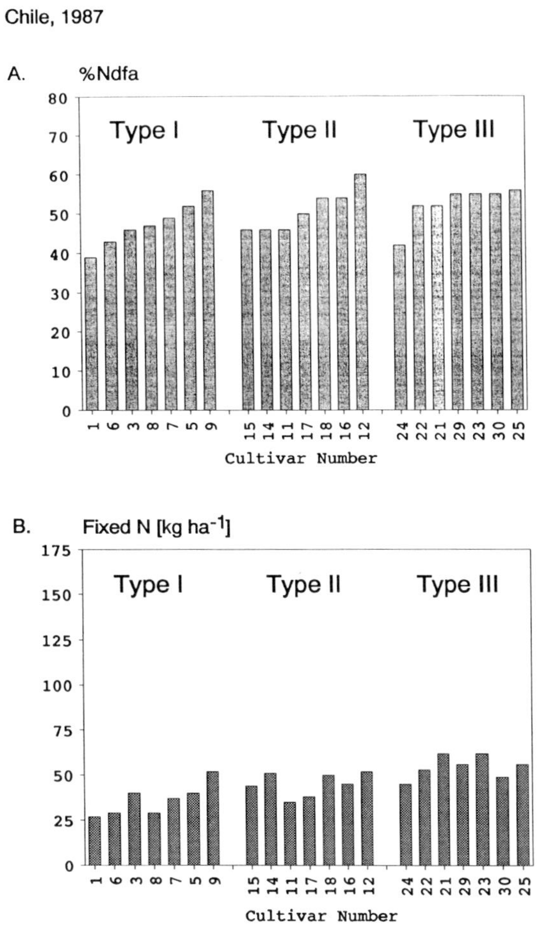

Fig. 4. (**A**) %N derived from atmosphere (%Ndfa) of common bean cultivars of types I, II and III tested under field conditions in Chile in 1987 by L. Longeri and (**B**) the amount of N_2 fixed by the same cultivars. The cultivars are listed in Table 3.

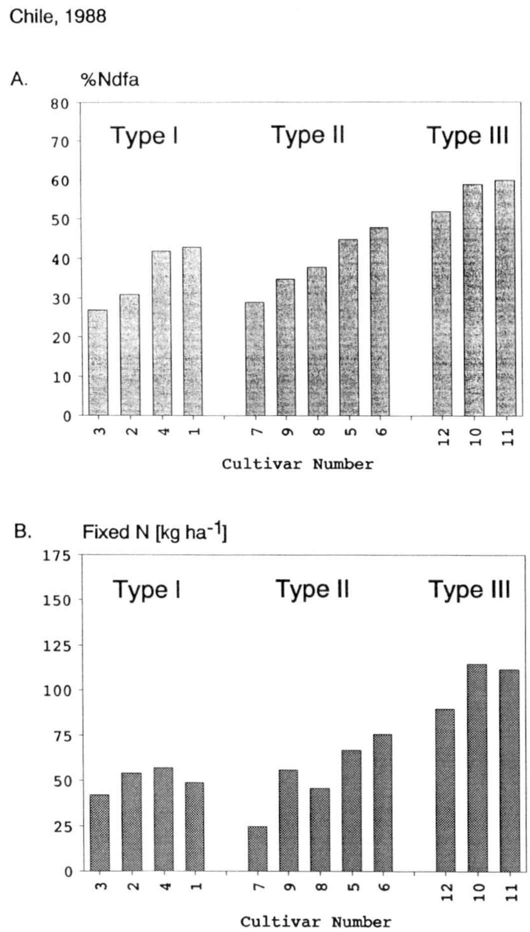

Fig. 5. (**A**) %N derived from atmosphere (%Ndfa) of common bean cultivars of types I, II and III tested under field conditions in Chile in 1988 by L. Longeri and (**B**) the amount of N_2 fixed by the same cultivars. The cultivars are listed in Table 3.

5A, B). The three lines with both the highest % Ndfa and most fixed N ha^{-1} all had Type III plant habit. On the average, type III plants fixed much more % Ndfa than types I and II, and 50 kg N ha^{-1} more than either types I or II which were similar.

Colombia

The experiment conducted at the Centro Internacional Agricultura Tropical (CIAT) in Colombia included nine bean lines and sorghum as the non-fixing reference crop (Table 4). The Ndfa ranged from 32 to 47%, with line A268 being the highest (Fig. 6A). Line A268 also fixed the most total N, 36 kg ha^{-1} (Fig. 6B). There was a strong positive correlation between % Ndfa and kg N ha^{-1} for the lines in this experiment.

Table 4. Common bean cultivars tested by Kipe-Nolt (Colombia)

No.	Cultivars
1	A 268
2	Riz 30
3	Riz 36
4	Riz 27
5	Riz 29
6	Bat 76
7	Bat 1297
8	Bat 332
9	Bat 1554
Ref.	Sorghum

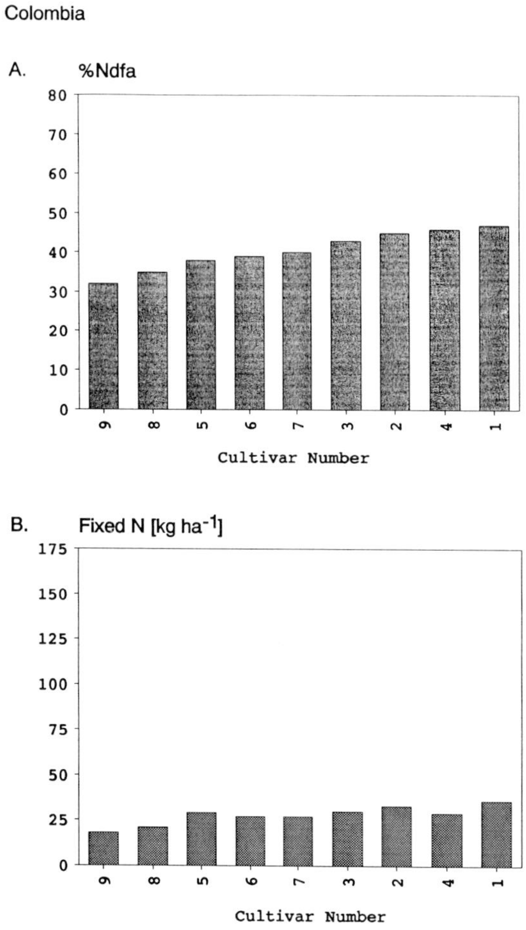

Fig. 6. (**A**) %N derived from atmosphere (%Ndfa) of common bean cultivars tested under field conditions in Colombia by J A Kipe-Nolt and (**B**) the amount of N_2 fixed by the same cultivars. The cultivars are listed in Table 4.

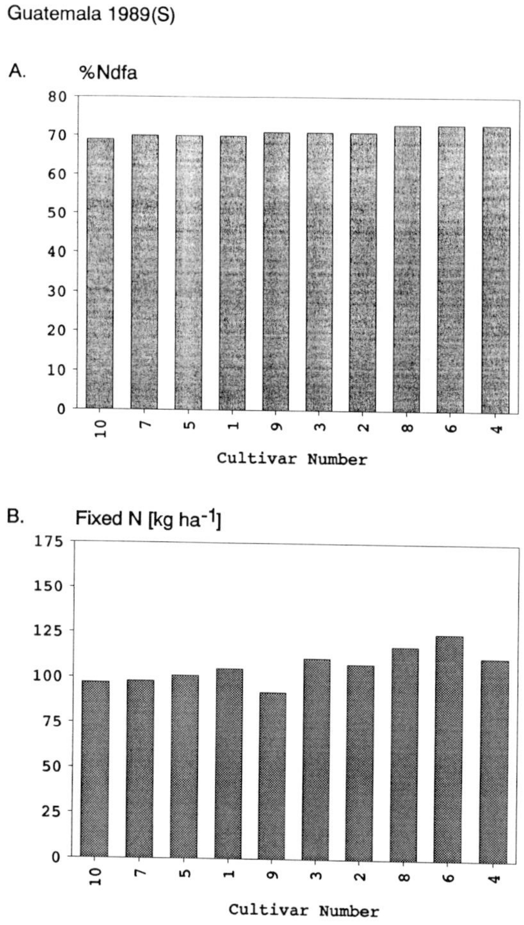

Fig. 7. (**A**) %N derived from atmosphere (%Ndfa) of common bean cultivars tested under field conditions (summer) in Guatemala by C A Sanabria and (**B**) the amount of N_2 fixed by the same cultivars. The cultivars are listed in Table 5.

Table 5. Common bean cultivars tested by Sanabria (Guatemala)

1989 summer		1989 winter	
No.	Cultivars	No.	Cultivars
1	ICTA – San Martin	1	ICTA – Tamazulapa
2	ICTA – Parramos	2	ICTA – Quetzal
3	ICTA – Quenack-Ché	3	Rabia de Gato
4	Compuesto Chimalteco 2	4	ICTA – Ostúa
5	ICTA – Texel	5	Turbo III
6	Line CH 86-48	6	Dor 364
7	Line CH 86-50	7	Puebla 152
8	Line CH 87-10	8	Wisconsin 21-58
9	Line CH 87-26	9	Negro Queretaro
10	Line CH 87-44	10	CU 85-15
Ref.	Non-nod Soybean, wheat	Ref.	Non-nod Soybean, Millet

Guatemala

The highest consistent values for % Ndfa and total fixed N among all the experiments were obtained for the 10 lines grown in Guatemala in the summer 1989 (Table 5). The range for Ndfa was from 69 to 73% and for total Ndfa, from 92 to 125 kg ha^{-1}. The best lines for % Ndfa and total fixed N were ICTA San Martín, ICTA Panamos and ICTA Quenack-Ché (Figs. 7A, B). The mean values for both traits were considerably lower for 10 other bean lines grown in the winter, 1989 (Figs. 8A, B), with the range being from 22 to 57% Ndfa. The range for total Ndfa was 12 to 42 kg N ha^{-1}.

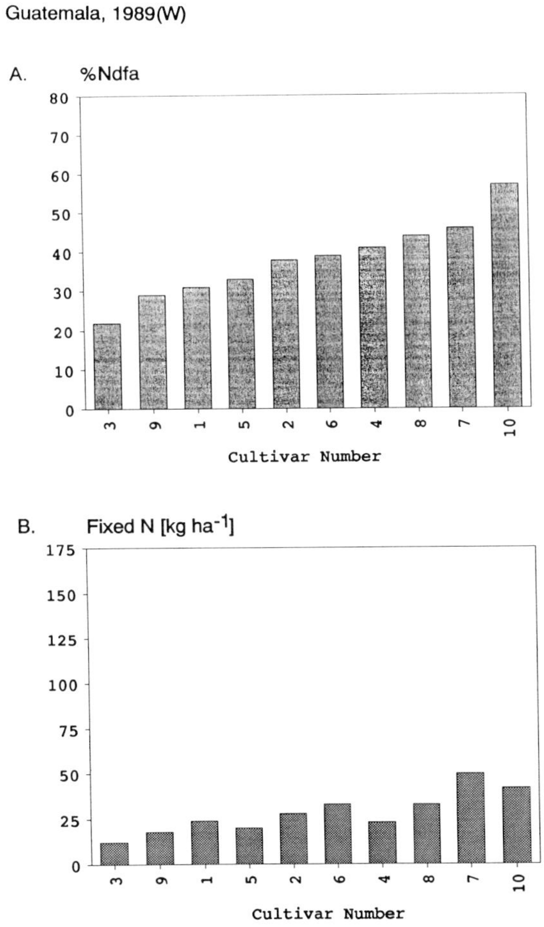

Fig. 8. (**A**) %N derived from atmosphere (%Ndfa) of common bean cultivars tested under field conditions (winter) in Guatemala by C A Sanabria and (**B**) the amount of N_2 fixed by the same cultivars. The cultivars are listed in Table 5.

Mexico

Eighteen cultivars with non-nod soybean as the non-fixing reference crop were grown at Colima, Mexico in the winter, 1988 (Table 6). The range

Table 6. Common bean cultivars tested by Cigales-Rivero at Colima, Mexico

No.	Cultivars
1	Alubia
2	Azufrado
3	Bayo Alteño
4	Bayo Gordo
5	Bayocel
6	Chícaro
7	Flor de Mayo
8	Jamapa
9	Negro Colima
10	Negro Huasteco
11	Negro Nayarit
12	Negro Poblano
13	Negro Tecomán
14	Ojo de Cabra
15	Pinto
16	Tempranillo
17	Zamorano
18	Honduras 35
Ref.	Non-nod soybean

Table 7. Common bean cultivars tested by Peña-Cabriales at Irapuato, Mexico

No.	Cultivars
6	Sataya 425
7	Bayocel
8	Flor de Mayo RMC
9	Apaseo 67
10	Negro Queretaro Criollo
11	Criollo Huejutla
12	Mantequilla de Calpán
13	Bayo Madero
14	Bayo. Zacatecas
15	Bayo Los llanos
16	Ojabra 400
17	Cacahuate Ags 19-3-2
18	Porillo Sintetico
19	Bayo Rio Grande
20	Wisconsin 21-58
21	WBR 22-3
22	Diacol Colima
23	Puebla 152
24	II-320-MRP-2-13-131
25	Compuesto Chimaltenango 2
Ref.	Sorghum

for % Ndfa was from 0 to 50% (Fig. 9A), and for total Ndfa from 0 to 70 kg ha^{-1} (Fig. 9B). The cultivars Azufrado, Negro Colima and Negro Poblano were the best for both measures of N_2 fixation.

Twenty bean lines with sorghum as the non-fixing reference crop were grown at Irapuato in 1987 (Table 10). The range for % Ndfa was from 5 to 58%, with the best lines being Criollo Huejutla, Negro Queretaro Criollo and Wisconsin 21–58 (Fig. 10A). The range for total Ndfa was from 7 to 108 kg ha^{-1}, with the best line being Wisconsin 21–58 (Fig. 10B).

Peru

Twenty bean lines with wheat as the non-fixing reference were grown in the winter season 1986 in Peru (Table 8). The values for % Ndfa ranged from 32–57% (Fig. 11A) and for total fixed N, from 19 to 59 kg ha^{-1} (Fig. 11B). For both measures of fixation, the best cultivars were Caballero, Caraota and Blanco. In the summer, 1988 season, % Ndfa was from 24 to 53% (Fig. 12A) and total Ndfa from 12 to 59% (Fig. 12B), with the best lines for % Ndfa being Bayo Normal, Canario G-62-2-6(26), and Bayo G-7.5-9.

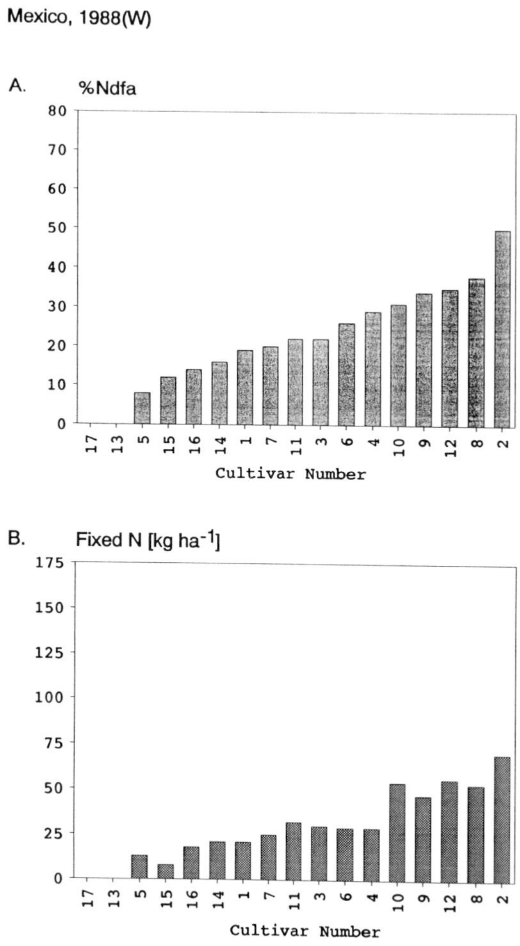

Fig. 9. (**A**) %N derived from atmosphere (%Ndfa) of common bean cultivars tested under field conditions in Mexico by M R Cigales-Rivero and (**B**) the amount of N_2 fixed by the same cultivars. The cultivars are listed in Table 6.

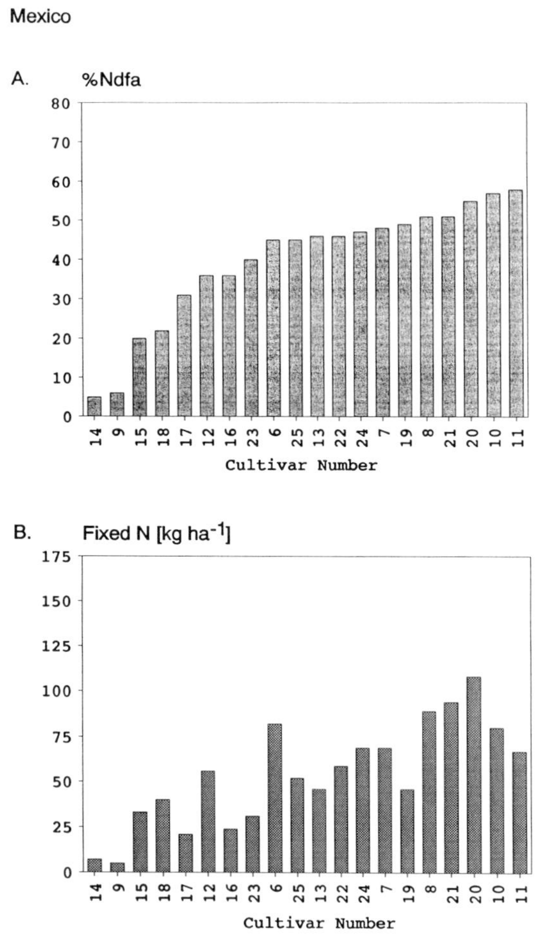

Fig. 10. (**A**) %N derived from atmosphere (%Ndfa) of common bean cultivars tested under field conditions in Mexico by J J Peña-Cabriales and (**B**) the amount of N_2 fixed by the same cultivars. The cultivars are listed in Table 7.

Table 8. Common bean cultivars tested by Manrique (Peru)

1986 winter		1988 summer	
No.	Cultivars	No.	Cultivars
1	Caballero	1	Caballero
2	Caraota	2	Bayo Normal
3	Blanco	3	Bayo G-10.0-5
4	Canario G-58-13(14)	4	Bayo G-7.5-9
5	Canario Divex	5	Canario G-62-2-6(26)
6	Canario G-62-13(32)	6	Canario G-62-4(27)
7	Canario G-62-4(27)	7	Negro
8	Panamito Grande	8	Caraota Venezolano
9	Canario G-187-2(35)	9	Canario G-252-17(31)
10	Bayo G-7.5-2	10	Canario Divex
11	Bayo G-7.5-9e	11	Canario G-62-13(32)
12	Canario G-25-18(4)	12	Tortola Diana
13	Negro	13	Panamito
14	Caraota Venezolano	14	Canario G-25-18(4)
15	Panamito	15	Canario G-58-13(14)
16	Red Kloud	16	Blanco
17	Canario G-62-2-6(26)	17	Bayo G-7.5-2
18	Bayo G-10.0-5	18	Canario G-187-2(35)
19	Bayo G-5.0-1e	19	Panamito Grande
20	Canario G-152-17(37)	20	Caraota Basileño
Ref.	Wheat	21	Bayo G-5.0-1e
		22	Red Kloud
		Ref.	Wheat

Table 9. Range of % and amount of N derived from atmosphere in *Phaseolus* cultivars tested by the participants in the programme

Location	Year	Range of % Ndfa	Range of fixation (kg ha^{-1})	Number of cultivars
Austria	1987	27–67	25–165	29
Brazil, Goiania	1987	12–25	4–12	17
Piracicaba	1987	19–53	11–53	7
Chile	1987	38–60	27–62	21
	1988	27–60	25–115	12
Colombia	1985	32–47	18–36	9
Guatemala	1989 (Summer)	69–73	92–125	10
	1989 (Winter)	22–57	12–50	10
Mexico, Irapuato	1987	5–58	7–108	20
Colima	1988 (Winter)	0–50	0–70	17
Peru	1986 (Winter)	24–56	15–59	20
	1988 (Summer)	13–56	7–81	22

Table 10. Nitrogen fixation of some commercial cultivars in comparison to the best ones tested in each trial

Location	Year	Commercial cultivar			Best cultivar		
		Cultivar	% Ndfa	Amount	Cultivar	% Ndfa	Amount
Brazil, Goiania	1987	Rio Tibagi	22	8	Honduras 35	25	12
Chile	1987	Tortola	44	50	Araucano	52	61
	1988	Tortola	52	90	Araucano	59	115
Guatemala	1989 (S)	ICTA – San Martin	70	105	Line CH 86-48	73	125
	1989 (W)	ICTA-Tamazulapa	31	24	CU85-15	57	42
Mexico, Irapuato		Flor de Mayo	51	89	Wisconsin 21-58	55	108
Colima	1988 (W)	Flor de Mayo	20	25	Azufrado	50	70
Peru	1987	Canario Divex	54	42	Caballero	57	59
	1988	Canario	47	31	Bayo Normal	53	40

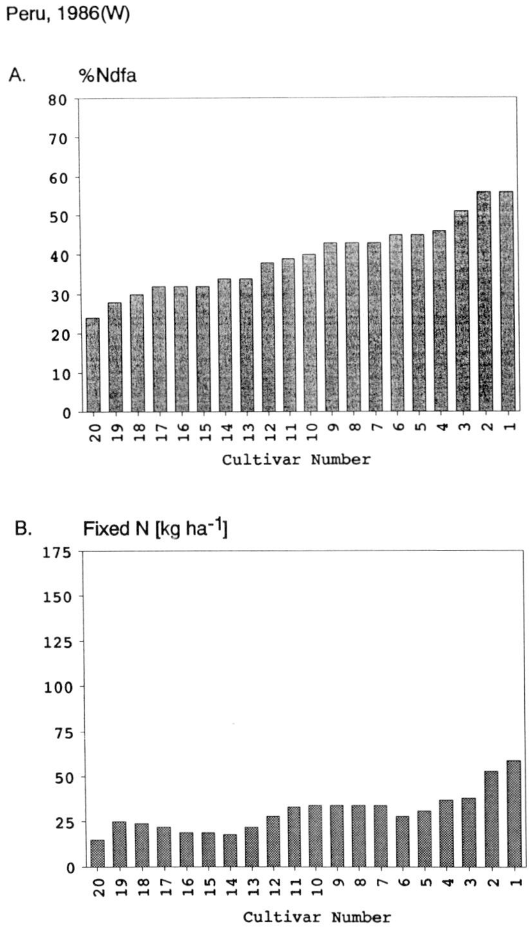

Fig. 11. (**A**) %N derived from atmosphere (%Ndfa) of common bean cultivars tested under field conditions (winter) in Peru by A Manrique and (**B**) the amount of N_2 fixed by the same cultivars. The cultivars are listed in Table 8.

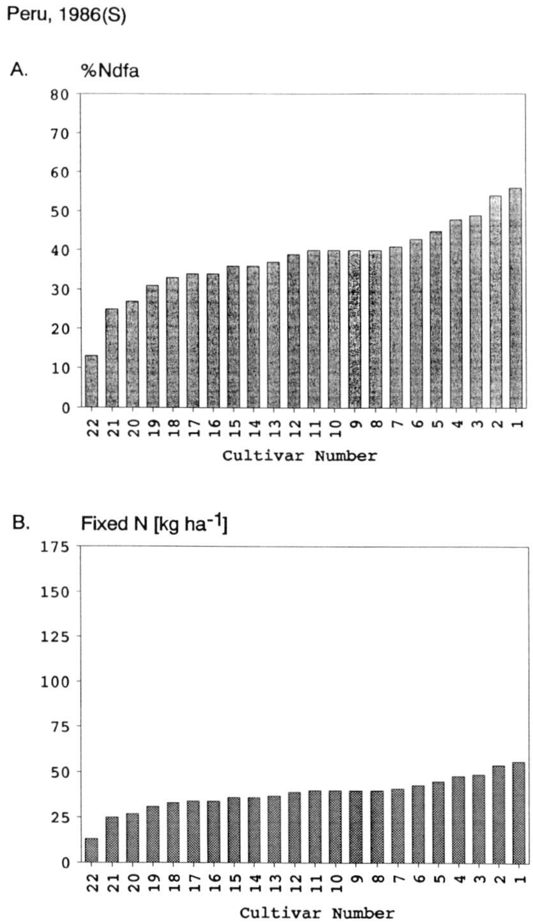

Fig. 12. (**A**) %N derived from atmosphere (%Ndfa) of common bean cultivars tested under field conditions (summer) in Peru by A Manrique and (**B**) the amount of N_2 fixed by the same cultivars. The cultivars are listed in Table 8.

Discussion

The data summarized in Table 9, illustrate the large differences that occurred among locations and seasons for both % Ndfa and total Ndfa. These differences reflect the sensitivity of the bean/*Rhizobium* symbiosis to many environmental factors that can have either an enhancing or reducing effect on N_2 fixation. The two experiments in Guatemala are interesting in several respects. Although both were done in 1989, the first was during the rainy season near Chimaltenago in the highlands and the second at a lower elevation near Jutiapa in eastern Guatemala. The bean lines grown were also different, but the mean amounts of soil N taken up by the lines in each experiment were quite similar (39.9 kg N ha^{-1} in 1989 summer vs. 40.6 kg N ha^{-1} in 1989 winter). The large difference between experiments were for N_2 fixation. In the summer, 1989 experiment, the mean values were 71% Ndfa and 107 kg Ndfa ha^{-1} vs. 38% Ndfa and 28 kg Ndfa ha^{-1} for the winter, 1989 experiment. Although the different lines undoubtedly are genetically different, dry conditions of the winter, 1989 experiment, were likely to be an important factor limiting N_2 fixation. Similarly, the cool, moist conditions for the summer, 1989 experiment, favoured optimal growth of the bean plants and high levels of N_2 fixation.

Fixation of lines grown at some other locations was quite low, i.e., Brazil. We speculate that these dry conditions and high temperatures contributed to low levels of fixation. In contrast to field results, preliminary results from a greenhouse experiment at the Seibersdorf laboratory in Austria, on the effect of different temperatures on N_2 fixation in soybean, faba bean and common bean did not support the hypothesis that common bean is more sensitive to high temperature (34°C) than other legumes (Hardarson, unpublished data).

There was great variability among genetically different lines in common experiments for both measures of N_2 fixation. Similar to other published results this study provides evidence for substantial genotypic variability. In some experiments up to 70% Ndfa was observed (e.g., summer, 1989, Guatemala), which is more than previously reported for common bean. In several other experiments, values of 50% Ndfa were observed consistently for a large number of cultivars. Also, the amounts of total plant Ndfa were consistently in the 50–60 kg N ha^{-1}, and in the Guatemala experiment of summer, 1989, averaged over 100 kg N ha^{-1} for the 10 bean lines.

The high values for N_2 fixation were observed on adapted cultivars and breeding lines when the environmental conditions were favourable. Some lines have the potential to be very efficient in supporting N_2 fixation, and where the high levels of fixation and favorable conditions were seen, seed yields were quite good. In the Guatemala (summer, 1989) experiment where fixation means were 71% Ndfa and 106 kg Ndfa, the mean seed yield of the 10 lines was 3.4 t ha^{-1} at 14% grain moisture. Low seed yields were prevalent where the total N from the soil, from fertilizer and from fixation were low. When high levels of N_2 fixation occurred, concomitant increases in total dry matter and seed yield were seen.

The results of the experiments in Chile provided additional evidence that in general, the more indeterminate and vining plant types fix more N_2 than those with determinate or bush habit. However, among the type I and II cultivars, date of maturity also influences N_2 fixation, with the later types fixing slightly more total Ndfa than earlier ones. There are important exceptions to the general pattern. For example, in 1987 Red Mexican, a type I cultivar, produced higher N_2 fixation than all the type II lines except Don Timoteo. Clearly, there is substantial variability and an opportunity to improve N_2 fixation in the different types by careful selection and breeding.

Some of the experimental lines tested in these experiments were better than the local cultivars included at each site (Table 10). With additional testing these might be recommended to farmers if other traits are suitable. Otherwise, lines with traits supporting high N_2 fixation can be used as parents, either to be improved for other important missing traits or as a donor for high N_2 fixation in crosses with otherwise good lines. Hybridization and selection have been done by some participants in this programme and some results are presented on the success of breeding common bean for enhanced nitrogen fixation. (Bliss, 1993).

Acknowledgements

Best thanks to Ms. H Axmann and the analytical group at the Seibersdorf Laboratory for isotope analyses and to Dr Anita Herndl-Silmbrod for making the graphics.

References

Bliss F A 1993 Breeding common bean for improved biological nitrogen fixation. Plant and Soil 152, 71–79.

Eastin E F 1978 Total nitrogen determination for plant material. Anal. Biochem. 85, 591–594.

FAO. Legume Inoculants and Their Use, 1984. FAO, Rome. 63 p.

Fiedler R and Proksch G 1975 The determination of ^{15}N by emission and mass spectrometry in biochemical analysis: A review. Anal. Chim. Acta. 78, 1–62.

Fried M and Middelboe V 1977 Measurement of amount of nitrogen fixed by a legume crop. Plant and Soil 47, 713–715.

Hardarson G 1990 Use of nuclear techniques in studies of soil-plant relationships. Training Course Series No. 2. IAEA. 223 p.

Hardarson G, Danso S K A and Zapata F 1987 Biological nitrogen fixation in field crops. *In* Handbook of Plant Science in Agriculture. Eds. B R Christie. pp 165–192. CRC Press Inc., Boca. Raton, FL.

Hardarson G and Lie T A (Eds.) 1984 Breeding Legumes for Enhanced Symbiotic Nitrogen Fixation. Martinus Nijhoff Publishers, Dordrecht. 166 p.

McAuliffe C, Chamblee D S, Uribe-Arago H and Woodhouse Jr. W W 1958 Influence of inorganic nitrogen on nitrogen fixation by legumes as revealed by N-15. Agron. J. 55, 334–337.

Plant and Soil **152**: 71–79, 1993.

PLSO BNF06

Breeding common bean for improved biological nitrogen fixation

F.A. BLISS
Department of Pomology, University of California, Davis, CA 95616-8683, USA

Key words: common bean, nitrogen fixation, *Phaseolus vulgaris* L.

Abstract

Common bean (*Phaseolus vulgaris* L.), which is an important food crop in the Americas, Africa and Asia, usually is thought to fix only small amounts of atmospheric nitrogen. However, field data indicate considerable genetic variability for total N_2 fixation and traits associated with fixation. Studies have shown that selection to increase N_2 fixation will be successful if: (1) discriminating traits (selection criteria) are measured precisely, (2) variability in germplasm is heritable, (3) selected parents are also agronomically suitable, (4) units of selection facilitate quantification of selection criteria, and (5) a breeding procedure that allows maximum genetic gain for N_2 fixation and recombination with essential agronomic traits is chosen. Breeding lines capable of fixing enough atmospheric N_2 to support seed yields of 1000–2000 kg ha^{-1} have been identified and new cultivars with high N_2 fixation potential are being released.

Introduction

Common bean (*Phaseolus vulgaris* L.), in symbiosis with *Rhizobium leguminosarum* biovar *phaseoli* has been considered a poor N_2-fixing grain legume. Both the total amount of nitrogen (N) derived from the atmosphere (total Ndfa) and the proportion of the plant N from the atmosphere (% Ndfa) in commonly-grown cultivars often are insufficient for producing economically attractive seed yields (e.g. 1,000–2,000 kg ha^{-1}). In contrast to the belief that beans are invariably poor N_2 fixers, there is evidence for genotypic variability for traits associated with N_2 fixation potential (e.g. Graham and Rosas, 1977; Park and Buttery, 1988, 1989); and for amount of N_2 fixed (e.g., McFerson, 1983; Rennie and Kemp, 1983; Pereira et al., 1989). Furthermore, selection has produced breeding lines able to fix high levels of N_2 (McFerson, 1983; Miranda and Bliss, 1991; St. Clair et al., 1988; St. Clair and Bliss, 1991), and the improved breeding lines are being tested for commercial use (Bliss et al., 1989). In field experiments some bean lines have been shown to fix enough N_2 to support seed yields of 1,000–2,000 kg ha^{-1}, when effective rhizobial populations are present either naturally or from inoculation, and there are no other major yield limiting factors.

Because of traditionally low returns, common bean often is produced on marginal soils. Other mineral elements besides N may be limiting, high temperatures and insufficient moisture often prevail, and production practices are less than optimal. The limitations to N_2 fixation imposed by the plant genotype can now be alleviated through breeding and selection of improved cultivars; therefore a 'production package' that is suitable for the small farmer using minimal purchased inputs should be designed and implemented. This package should include; an adapted cultivar capable of high N_2 fixation, assurance of an effective and efficient rhizobial population through proper inoculation, a source of high quality, disease-free seed, and management practices that are acceptable to the farmer for the prevailing agroecological system.

Effective selection for increased N_2 fixation

Bliss (1991) pointed out that effective selection for increased N_2 fixation in grain legumes depends on the following factors: (1) choice of traits as selection criteria that can be measured precisely and economically, while providing for discrimination between superior and inferior selection units; (2) variability in legume germplasm and heritability of differences for either plant N derived from the atmosphere (Ndfa) or traits indicative of fixation potential; (3) identification of genetically diverse parents that are also suitable agronomically; (4) choice of selection units (i.e., individual plants or families) that facilitate precise quantification of selection criteria and allow production of progeny from selected plants; and (5) use of a breeding procedure (e.g. mass selection, family selection) that provides maximum genetic gain per year for increased N_2 fixation and recombination with essential agronomic traits. If improved N_2 fixation potential is to be realized, selection must be practiced under soil N conditions that allow discrimination between high- and low-fixing lines. Field trials should be carried out without application of large amounts of N fertilizer to facilitate selection and simulate farming conditions where low purchased inputs are common. The practice of "on-farm" testing in environments that reflect real-life conditions should be expanded.

Methods of estimating N_2 fixation and choice of traits as selection criteria

Several methods can be used to estimate N_2 fixation potential and evaluate traits related to N_2 fixation. The acetylene reduction (AR) assay has been used extensively to estimate the potential for N_2 fixation at a specific time point (Burris and Wilson, 1957; McFerson, 1983; Turner and Gibson, 1980). Although AR has been useful for identifying effective plant-rhizobial combinations and the potential of plant genotypes for N_2 fixation, it should not be used for quantitative estimates of N_2 fixation (see other chapters in this book). Furthermore, since most AR assays utilize mainly the crown roots and central root core, fixation by lateral root nodules may be underestimated. Studies by Wolyn et al. (1989) in common bean and Hardarson et al. (1989) in soybean have shown that fixation by lateral root nodules contribute substantially to total N_2 fixation. Contrary to suggestions, that in grain legumes N_2 fixation declines at the onset of seed-fill, N_2 fixation in some common bean lines continues actively during seed fill, with much of the total N accumulation occurring after the R3 stage (Table 1) (Attewell and Bliss, 1985; Peña-Cabriales et al., 1993). Other estimates of N_2 fixation potential can be based on nodulation traits such as nodule mass, nodule number, nodulation score, etc. (Rosas and Bliss, 1986). Plant biomass, total plant N and grain yield are useful indicators of N_2 fixation for plants grown on minimal or no-N media.

The application of ^{15}N isotopic-labelled fertilizer to actively fixing legumes and a non-fixing standard crop allows quantitative estimation of the amounts and proportions of plant N coming from the soil (Ndfs), fertilizer (Ndff) and the atmosphere (Ndfa). The accuracy of this method for determining Ndfa depends on proper use of a suitable non-fixing reference crop (see other chapters in this book). Non-nodulating isolines of soybean have been used satisfactorily where they are well-adapted to the prevailing day-length. The availability of a non-nodulating, adapted bean line would be quite useful, but to date, none has been developed. Several other crops (e.g. barley, wheat, maize, sorghum) have been used with mixed results (see other chapters in this book). However it should though be mentioned that relative screening of common bean lines can be done without a reference crop.

The total N difference method can be used to estimate N_2 fixation when plants are grown on low-N soil and when either minimal or no N was applied. Either a non-fixing reference crop or the bean line with the least total accumulated N can be used as the minimal-fixing reference (Miranda and Bliss, 1991). The Kjeldahl method or infrared spectrophotometry calibrated against Kjeldahl N (Mundel and Schaalji, 1988) can be

Table 1. Total nitrogen accumulation in plants of progeny lines and parents of population 24. Field, Wisc., 1984

Line	Total plant N at			
	R1 (mg plant^{-1})	R3	R7	R3/R7
24-40	36	289	552	0.52
24-55	32	322	668	0.48
24-21	50	393	1045	0.38
24-17	49	172	1068	0.16
24-5	54	466	1149	0.41
24-12	57	198	1388	0.14
24-4	57	282	1396	0.20
24-19	37	265	1471	0.18
Sanilac	45	207	591	0.35
Puebla 152	68	646	1429	0.45

Source: Attewell and Bliss, 1985.

used to measure total N in the different plant tissues.

Various selection criteria can be used effectively, as long as it recognized that the primary trait for improvement is the amount and proportion of plant N derived from fixation. Therefore, selection criteria other than Ndfa must be viewed as indirect selection, and Ndfa as a correlated response. The selection method chosen often will reflect the capabilities of the experimenter to conduct the work effectively. However, a method chosen for simplicity, ease of application and low cost, but which produces little genetic gain for the trait of interest (i.e. N_2 fixation) is of limited value and should be avoided. Ultimately the seed yield resulting from the N present in the plant is important.

Variability for N_2 fixation and related traits

There is substantial genotypic variation for N_2 fixation and traits related to fixation in common bean germplasm (see other chapters in this book). Although there are few published estimates of heritability for these traits, usually environmental and genotype × environment effects are large, resulting in low to moderate heritability, especially when measured on a single plant basis. Therefore, greater selection efficiency can be achieved if selection is based on family mean rather than individual value as the unit of selection. The evaluation of families rather than individual plant allows replication for reducing the effects of non-genetic (environmental) factors that obscure genotypic value and for improving the precision of selection.

Most methods of estimating N_2 fixation, i.e., acetylene reduction, ^{15}N isotopic labelling, total plant N, are plant destructive. Therefore, both destructive plant sampling for analysis and saving the seed of single plant selections are not possible. The evaluation of families allows sampling of a portion of the family for estimation of N_2 fixation and other traits if desired, then saving seed from the remaining plants in the selected lines (Wynne et al., 1987).

In common bean, production of inbred families following hybridization of an adapted cultivar with a high N_2-fixing parent can be accomplished using either single seed descent (SSD) or the inbred backcross line (IBL) method in which one or two backcrosses is made to the adapted cultivar following the initial cross to the high fixing donor line. After backcrossing, selfing for one or two generations using SSD is practiced to produce near-homozygous IBL's for replicated testing (McFerson et al., 1982).

Breeding methods

Several factors influence the choice of breeding methods most suitable for improving N_2 fixation in common bean. Usually, there are distinct regional preferences for the commonly-grown cultivars, which relate to cropping system, day length, plant type and seed characteristics. In addition to being grown in many diverse areas, there are numerous consumer preferences based on seed shape and color, cooking quality, plant type, adaptability, etc. No single cultivar enjoys exclusive, wide-spread production; instead many different bean cultivars are grown worldwide or even in one region of a single country. Therefore, to meet these varied demands, breeding for improved N_2 fixation must be incorporated into a bean improvement program as a standard practice and be given a priority equivalent to other objectives such as yield improvement, suitable plant type, disease and insect resistance, adaptability, etc.

Practically speaking, the breeding method used must allow for easy recovery of a desirable genotype(s) when N_2 fixation is being improved. Recovery of elite genotypes may be enhanced by intercrossing parents with similar phenotypes, but often the lines with high N_2-fixation potential (e.g. Puebla 152) are unlike the standard cultivar being improved (e.g. cv. Sanilac). If the phenotypes of the parents are similar, modified pedigree selection among inbred lines developed by SDD or recurrent mass selection among families produced by intermating the phenotypically similar parents can be used effectively. If the donor line having increased N_2 fixation is quite different from the standard cultivar, backcrossing is an effective means of recovering the recurrent parent genotype while maintaining adequate variability for the trait being improved. The production of inbred backcross lines combines the desirable features of backcrossing and the production of inbred lines using SSD (Fig. 1). Since N_2 fixation is quantitatively expressed, single plant heritability is low, and methods for assaying N_2 fixation are plant destructive, the inbred backcross line (IBL) method facilitates breeding for improved N_2 fixation by allowing for family selection based on replication.

Family selection also allows easier evaluation of experimental units for multiple traits since portions of the same replicate (e.g. row plot) or different replicates of the same family can be evaluated for different traits. The improvement of any trait in common bean should allow also for simultaneous evaluation of yield to ensure

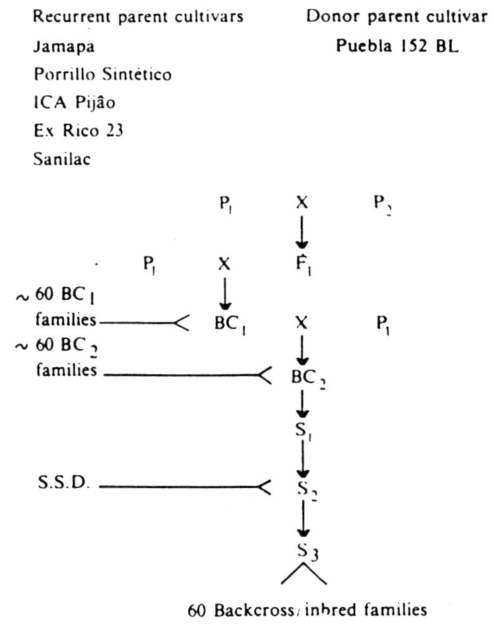

Fig. 1. The backcross inbred method of population development. Source: McFerson et al., 1982.

either that yield is improved or at least maintained, because yields of most current cultivars are somewhat low and improvement usually is of a high priority in breeding programs. The symbiotic nature of the host-rhizobial association requires that evaluation of the effectiveness of either natural populations or inoculant of *R. phaseoli* be done simultaneously with plant genotype improvement. The breeding method should allow easy evaluation of host-inoculant combinations in multiple locations and years. Use of family mean performance is well suited for these multiple evaluations.

Realized plant improvement from selection

There are only a few published reports where the results of hybridizing selected parents, selecting the best families in variable offspring populations and evaluating the selections under field conditions have been studied. In those cases, the results support the predictions that genetic improvement through breeding for N_2 fixation in common bean is feasible. There are no empirical data to suggest that increasing N_2 fixation of adapted bean lines is accompanied by unfavorable effects on other desirable traits (e.g., yield) if suitable breeding methods that allow evaluation of multiple traits are used.

Graham (1981) and Graham and Rosas (1977) reported that, based on conclusions from acetylene reduction assays, plants with a larger photosynthetic area fix more N_2 than those with less capability. That finding supported the conclusion that type IV (indeterminate, climbing) plants fixed more N_2 than type III, and that type I plants with determinate bush habit were invariably poor N_2 fixers. While the latter often is true, there are notable exceptions, especially within snapbean cultivars. In trials on low-N, sandy soils in Wisconsin, the cv. Lake Shasta was found to fix up to 70–80 kg ha^{-1} during the 60–70 day growing period based on difference method estimates using non-nod soybeans as the non-fixing crop (F.A. Bliss and K.A. Kmiecik, unpublished data). The results of Graham and Rosas (1977) also led people to conclude that since many existing type I lines are poor N_2 fixers and that indeterminate types (i.e., types II, III and IV) with greater photosynthetic capacity were capable of fixing more N_2, breeding for improved N_2 fixation in type I and II bean plants would be of little value. This has not been the case, since photosynthetic capacity is only one of several (many) plant factors affecting N_2 fixation.

The Mexican landrace cultivar Puebla 152 (black-seeded) which has a type III growth habit has been identified as a superior high N_2 fixing line based on AR assays. Puebla 152 has been used widely as a parent in crosses with type I and II cultivars. Superior selections from among inbred backcross progeny lines from crosses with Sanilac (type I, white-seeded), Porrillo Sintetico (type II, black-seeded), Jamapa (type II, black-seeded), ICA Pijao (type II, black-seeded) and Ex Rico 23 (type II, white-seeded) have been identified (McFerson, 1983). Several traits have been evaluated including total plant N, total and % Ndfa in the plant, total and % Ndfa in the seed, and seed yield. Various methods have been used to estimate N_2 fixation, including ^{15}N isotopic estimation using labelled fertilizers, the difference method with non-nod soybean as the non-fixing standard, and indirect measures based on total plant and seed N, and nodulation traits such as nodule weight mass, number and score. In all segregating populations, recombinant lines have been recovered with increased N_2 fixation or fixation potential and acceptable seed yield, and other agronomic traits similar to or exceeding the standard parent. Sometimes, as with other traits, further improvement is necessary because important traits (e.g. disease resistance) may be lacking.

The white navy bean cv. Sanilac was grown widely in the U.S., especially in Michigan and other north central states, until recently. Although possessing many desirable traits, it produces modest yields and is a poor N_2-fixing plant. The population of inbred backcross lines resulting from the cross of Sanilac × Puebla 152 has been studied extensively (McFerson et al., 1983; and McFerson, 1983, St. Clair et al., 1988). Promising selections based on AR assay, seed yield, plant type and days to harvest were made in field trials on low-N sandy soils. St. Claire et al. (1988) reported that some BC_2S_3 lines with agronomic traits similar to Sanilac (e.g., 24–21, 24–55) had more total Ndfa and % Ndfa than the recurrent parent but less than the donor parent, Puebla 152 (Table 2). Among S_2 families produced from intercrosses among the best BC_2S_4 lines there were superior families for high N_2 fixation that also had satisfactory seed yield, seed traits, early maturing and other good agronomic traits (St. Clair and Bliss, 1991). The total Ndfa and % Ndfa of the best families exceeded that of the recurrent parent Sanilac and were similar to Puebla 152 (Table 3). Indeterminate plant type was not an essential trait for expressing high N_2 fixation, although a larger leaf area presumably can contribute more photosynthate for plant growth, N_2 fixation and seed yield.

ICA Pijao is an outstanding type II black bean cultivar bred for subtropical conditions (i.e. CIAT, 1986). However, depending on the situation, it has been found to fix variable amounts of N_2. A population of BC_2S_2 lines derived from the cross of ICA Pijao × Puebla 152 showed substantial segregation for numerous important traits, including seed yield, plant type, maturity, N_2 fixation and growth under low phosphorous levels (Pereira and Bliss, 1989). Lines resulting

Table 2. Estimates of N_2 fixation, N yield per plant, and % Ndfa for common bean inbred backcross lines and parents at the R9 growth stage. Hancock, WI, 1985. Line mean values within each ^{15}N experiment (En = ^{15}N-enriched, Dp = ^{15}N-depleted)

Line	N yield		N_2 fixed		Ndfa (%)	
	En	Dp	En	Dp	En	Dp
	(mg N/plant)					
21-16	986	1158	584	450	59.2	38.4
21-38	1074	919	649	337	60.8	36.8
21-43	985	1076	510	471	52.0	43.1
21-58	1022	1029	634	302	54.2	29.2
Porrillo Sintetico	1163	1230	643	550	54.3	44.9
Puebla 152	1053	1392	661	688	62.5	49.3
Sanilac	629	848	72	19	9.8	3.1
24-17	1010	1136	507	442	48.0	39.1
24-21	933	1041	391	222	40.6	21.2
24-48	784	866	329	136	40.9	16.3
24-55	741	931	296	288	37.5	31.3
24-65	933	1077	417	242	44.8	24.3
LSD (0.05)	268	243	224	167	14.4	12.9
r_S †	0.85**		0.61*		0.74**	

*, ** Significance at the 0.05 and 0.01 levels, respectively.
† r_S = Spearman rank correlation coefficients for line ranks between En and Dp experiments.
Source: St. Clair et al., 1988.

from this cross (i.e., Reg. No. GP-71-PI520603; GP-72-PI520604; GP-73-PI520605; GP-74-PI520606; GP-75-PI520607) which were selected and evaluated (e.g., CIAT, 1987) in the U.S. and Brazil have been released as improved germplasm having improved N_2 fixation ability (Bliss et al., 1989). The selected line WBR 22–34 was compared to the widely-grown standard cv. Rio Tibagi and to Negro Argel which has been found to have better nodulation. WBR 22–34 fixed more total Ndfa and % Ndfa and was higher yielding than these two Brazilian black bean cultivars.

The five released lines (Bliss et al., 1989) as well as other breeding lines selected at CNPAF/EMBRAPA for high N_2 fixation and important agronomic traits (e.g. Honduras 35) are being evaluated in Brazilian state trials. Although moderate to high yields usually are reported under low-N soil conditions, their utility as recommended cultivars will depend also on other traits such as several disease reactions, i.e., bean rust, bean common mosaic virus and anthracnose. For regions where these diseases are limiting factors to production, breeding is underway to combine resistance and high N_2 fixing ability into adapted breeding lines (Pereira, P.A.A., CNPAF/EMBRAPA, personal communication).

Because common beans of many different seed types are grown in Brazil (e.g., carioca, mulatinho, rosinha, etc.), the incorporation of genes for increased N_2 fixation into white navy and black-seeded cultivars is viewed as a model system that can be extended to other materials. The concept of selection for improved N_2 fixation has been incorporated into the objectives of the bean improvement program at CNPAF/EMBRAPA in a manner similar to that for other important traits, i.e., selection for yield, disease resistance, etc. (Pereira, P.A.A.; person communication). Currently, priority is given to identifying high N_2 fixing parents, selecting for high N_2 fixation in segregating populations and evaluating promising selections made for high yield, resistance and good adaptability for N_2 fixation as well.

Miranda and Bliss (1991) reported genotypic variability for total seed N in populations of F_3 lines resulting from crosses of different breeding lines and cultivars to a single testor chosen for N_2 fixation ability. Family mean selection among plants grown on a low-N soil resulted in lines

Table 3. Mean values of five traits for 15 F, families (i.e., 10% of the population) selected for high N_2 fixation, the four parental IB lines, the recurrent parent 'Sanilac' and the donor parent 'Puebla 152', Hancock, WI, 1985

Entries	At mid-podfill (R7)				At maturity (R9)
	Ndfa (%)	N_2 fixed (mg/plant)	Shoot N (mg/plant)	Shoot dry wt (g/plant)	Seed yield (g/plant)
(17 × 48)-22	60.8	1028.8	1680	58.6	31.5
(17 × 48)-17	52.7	911.8	1650	59.2	32.5
(48 × 65)-22	54.2	910.2	1678	67.1	26.3
(17 × 21)-09	53.3	791.7	1465	54.6	28.7
(17 × 65)-02	55.5	779.3	1384	52.6	29.8
(17 × 21)-04	50.7	756.4	1484	58.0	24.1
(17 × 65)-06	49.3	751.0	1519	58.3	30.6
(17 × 48)-18	57.9	739.1	1270	46.4	24.4
(48 × 65)-23	46.4	736.9	1576	64.0	22.3
(17 × 65)-16	51.2	721.4	1416	54.8	27.8
(17 × 21)-06	47.3	718.2	1485	50.4	23.1
(17 × 65)-25	43.4	707.2	1511	63.0	38.6
(17 × 48)-19	42.5	675.2	1565	59.0	32.5
(17 × 65)-20	44.0	672.4	1518	58.5	25.9
(17 × 48)-16	49.9	667.0	1359	46.9	23.8
(17 × 21)-06	47.3	718.2	1485	50.4	23.1
(17 × 65)-25	43.4	702.2	1511	63.0	38.6
(17 × 48)-19	42.5	675.2	1565	59.0	32.5
(17 × 65)-20	44.0	672.4	1518	58.5	25.9
(17 × 48)-16	49.9	667.0	1359	46.9	23.8
24-17	45.9	573.4	1267	48.1	31.8
24-21	37.2	414.3	1142	46.5	23.3
24-48	37.3	426.2	1146	43.4	18.0
24-65	36.2	504.4	1348	59.9	29.8
Puebla 152	50.4	589.2	1128	46.9	42.3
Sanilac	19.0	142.5	729	31.8	16.0
$LSD_{0.05}$*	12.8	336.5	487	16.2	7.2

* LSD mean comparisons at ($p < 0.05$).
Source: St. Clair and Bliss, 1991.

with increased total plant N and seed N accumulation (Fig. 2). Since no fertilizer N was applied and non-nodulating soybeans were N deficient, it was concluded that increased seed N of the selections compared to the parents, resulted from more N_2 fixation.

This method of indirect selection for increased N_2 fixation can be practiced successfully where there are low levels of soil N and no fertilizer is added. Simultaneous improvement for seed yield is also possible depending on the amount of genotypic variability in the segregating populations.

If breeding for increased N_2 fixation is to receive high priority along with other improvement objectives, there must be a clear recognition that: 1) high N_2 fixation is desirable, 2) adequate genetic variability is available, 3) selection for increased N_2 fixation is feasible using suitable methods, and 4) practical methods are available for identifying high N_2 fixing lines. If it is not viewed as such, N_2 fixation probably will not be improved. When other traits are given higher priority and breeding lines are evaluated on N-fertile soils, the farmer growing beans on infertile soils with minimal N input is not likely to receive new cultivars able to derive much of the required nitrogen from N_2 fixation.

The soil N level on which segregating lines are evaluated is crucial to successful selection for increase N_2 fixation. Where inherent N fertility is high or substantial fertilizer N has been applied it is impossible to discriminate among lines for differences in N_2 fixation without the use of ^{15}N isotope techniques. The technical demands and costs of these techniques may preclude use on a large scale to evaluate numerous breeding lines. Furthermore, evaluation under high-N condi-

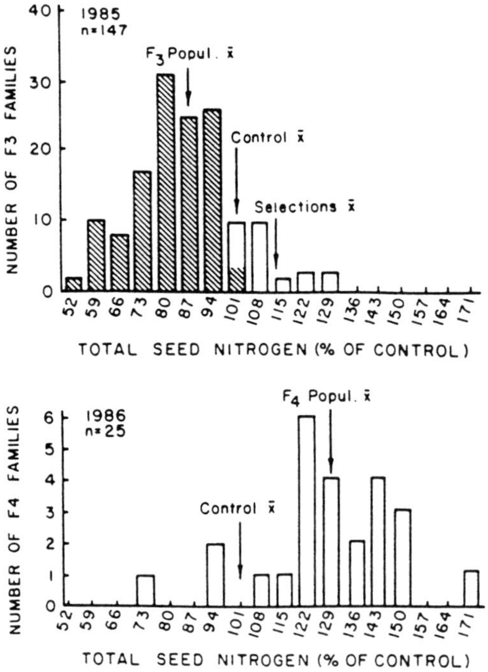

Fig. 2. Frequency distribution based on percentage of control of a population of F_3 half-sib families in 1985 and F_4 families resulting from the selected F_3's in 1986 for total seed nitrogen. Source: Miranda and Bliss, 1991.

tions will allow identification only of high N_2-fixing lines insensitive to combined N (if they exist); and those capable of fixing under low-N conditions will not be retained. Inclusion of an adapted, non-nodulating legume along with the breeding lines being tested provides a valuable visible means of evaluating the relative level of N coming from sources other than fixation.

Despite knowledge that selection among breeding lines for high N_2 fixation has been practiced and that the farmer is likely to grow beans on low-N soils, standard yield trials grown at experiment stations to evaluate lines prior to release, are often conducted with the addition of fertilizer N at high levels not likely to be used by the farmer. There is a belief among many researchers that a 'good trial' is a high yielding trial with a low coefficient of variation (CV). When high yields are achieved by adding N fertilizer it is difficult to evaluate promising lines selected for increased N_2 fixation, and these conditions are not likely to be encountered on the farms.

Future research priorities

Since ample genotypic variability for N_2 fixation in common bean has been identified in the primary gene pool, more effort in bean improvement programs should be placed on selection for increased N_2 fixation under representative field conditions and involving improved inoculant when possible. Accurate assay methods are available to allow discrimination between high- and low-N_2 fixing lines. The family mean rather than individual phenotype should be the unit of selection to allow replication and accommodate plant destructive methods used in the assay for increased N_2 fixation. Production of segregating families from planned crosses can follow either the modified pedigree selection method where lines are derived by single seed decent or the inbred backcross line method where near homozygous lines resembling the recurrent parent are produced. Evaluation of segregating progenies and of advanced lines should be on low-N soils representative of those that prevail in the target production area.

Both the rhizobia and plant genotypes often are sensitive to high levels of combined N, with N_2 fixation often being suppressed by added fertilizer N. There is evidence for both natural variants and induced mutants (Park and Buttery, 1989; St. Clair et al., 1988) of common bean able to tolerate combined N. This characteristic along with the capacity to fix large amounts of N_2 should be bred into high-quality, well-adapted lines with acceptable seed traits.

Despite extensive observation of many plant materials, no natural non-nodulating mutants of *Phaseolus vulgaris* have been identified. Several induced mutants incapable of forming effective nodules have been described, but non-nodulating lines of the major-classes, e.g., black, red kidney, carioca, etc., have not yet been developed. Although, of limited practical value, well-adapted non-nodulating lines would be useful in field selection and evaluation experiments to verify the level of soil N and to use as the non-fixing reference crop when estimates of N_2 fixation using the ^{15}N isotopic and total N difference methods are desired.

Few plant genes controlling expression of the different aspects of N_2 fixation in beans have been identified. Efforts should be made to elucidate plant genetic control more clearly and once identified, to place the genes on a genetic map along with other traits. If linkage with molecular markers can be established, marker-facilitated selection might be used to complement direct and indirect selection for increased N_2 fixation.

References

Attewell J and Bliss F A 1985 Host plant characteristics of common bean lines selected using indirect measures of N_2 fixation. *In* Nitrogen Fixation Research Progress. Eds. HJ Evans, PJ Bottomly and WE Newton. pp 3–9. Martinus Nijhoff Publ., Dordrecht.

Bliss F A 1991 Breeding plants for enhanced beneficial interactions with soil microorganisms. *In* Plant Breeding in the 1990s. Eds. H T Stalken and J Murphey. pp 251–273. C.A.B. International, Wellingford, UK.

Bliss F A, Pereira P A A, Araujo R S, Henson R A, Kmiecik K A, McFerson J R, Teixeira M G and da Silva C C 1989 Registration of five high nitrogen fixing common bean germplasm lines. Crop Sci. 29, 240–241.

Burris, R H and Wilson P W 1957 Methods for measurement of nitrogen fixation. *In* Methods in Enzymology. Eds. S P Colwick and N O Kaplan. Vol. 4. pp 355–366. Academic Press, New York, NY.

CIAT 1986 Annual Report. Centro International de Agricultura Tropicale, Cali, Columbia.

CIAT 1987 Annual Report. Centro International de Agricultura Tropicale, Cali, Columbia.

Graham P H 1981 Some problems of nodulation and symbiotic fixation in *Phaseolus vulgaris* L.: A review. Field Crops Res. 4, 93–112.

Graham P H and Rosas J H 1977 Growth and development of indeterminate bush and climbing cultivars of *Phaseolus vulgaris* L. inoculated with *Rhizobium*. J. Agric. Sci. Camb 88, 503–508.

Hardarson G, Golbs M and Danso S K A 1989 Nitrogen fixation in soybean [*Glycine max* (L.) Merrill] as affected by nodulation patterns. Soil Biol. Biochem. 21, 783–787.

McFerson J R 1983 Genetic and Breeding Studies of Dinitrogen Fixation in Common Bean, *Phaseolus vulgaris* L. Ph.D. Diss., University of Wisconsin, Madison: (Diss. Abst. Intern. DA8313188, Ann Arbor, MI).

McFerson J R, Bliss F A and Rosas J C 1982 Selection for enhanced nitrogen fixation in common bean, *Phaseolus vulgaris* L. *In* Biological Nitrogen Fixation Technology for Tropical Agriculture. Eds. P H Graham and S Harris. pp 39–44. CIAT, Cali, Columbia.

Miranda B D and Bliss F A 1991 Selection for increased seed nitrogen accumulation in common bean: Implications for improving Dinitrogen fixation and seed yield. Plant Breeding 106, 301–311.

Mundel H H and Schaal J I 1988 Use of near infrared reflectance spectroscopy to screen soybean lines for plant nitrogen. Crop Sci 28, 157–162.

Park S J and Buttery B R 1988 Nodulation mutants of white bean (*Phaseolus vulgaris* L.) induced by ethyl-methane sulphonate. Can. J. Plant Sci. 68, 199–202.

Park S J and Buttery B R 1989 Identification and characterization of common bean (*Phaseolus vulgaris* L.) lines well-nodulated in the presence of high nitrate. Plant and Soil 119, 237–244.

Peña-Cabriales J J, Grageda-Cabrera O A, Kola V and Hardarson G 1993 Time course of N_2 fixation in common bean (*Phaseolus vulgaris* L.). Plant and Soil 152, 115–121.

Pereira P A A and Bliss F A 1989 Selection of common bean (*Phaseolus vulgaris* L.) for N_2 fixation at different levels of available phosphorus under field and environmentally controlled conditions. Plant and Soil 115, 75–82.

Pereira P A A, Burris R H and Bliss F A 1989 ^{15}N-determined dinitrogen fixation potential of genetically diverse bean lines (*Phaseolus vulgaris* L.). Plant and Soil 120, 171–179.

Rennie R J and Kamp G A 1983 N_2 fixation in field beans (*Phaseolus vulgaris* L.) quantified by ^{15}N isotope dilution. II. Effect of cultivars of beans. Agron. J. 75, 645–649.

Rosas J C and Bliss F A 1986 Host plant traits associated with estimates of nodulation and nitrogen fixation in common bean. HortSci. 21, 287–289.

St. Clair D A and Bliss F A 1991 Intrapopulation recombination for ^{15}N-determined dinitrogen fixation ability in common bean. Plant Breeding 106, 215–225.

St. Clair D A, Wolyn D J, DuBois J, Burris R H and Bliss F A 1988 A field comparison of N_2 fixation determined with ^{15}N-enriched ammonium sulfate in selected inbred backcross lines of common bean. Crop Sci. 28, 773–778.

Turner G L and Gibson A H 1980 Measurement of nitrogen fixation by indirect means. *In* Methods for Evaluating Biological Nitrogen Fixation. Ed. FJ Bergersen. pp 111–138. Wiley, New York.

Wolyn D J, Attewell J, Ludden PW and Bliss F A 1989 Indirect measures of N_2 fixation in common bean (*Phaseolus vulgaris* L.) under field conditions. The role of lateral root nodules. Plant and Soil 113, 181–187.

Wynne J C, Bliss F A and Rosas J C 1987 Principles and practices of field designs to evaluate symbiotic nitrogen fixation. *In* Symbiotic Nitrogen Fixation Technology. Ed. G H Elkan. pp 371–389. Marcel Dekker, Inc., New York.

Plant and Soil **152**: 81–85, 1993.

PLSO BNF07

Mutation breeding of grain legumes

A. MICKE
Plant Breeding and Genetics Section, Joint FAO/IAEA Division, Vienna, Austria

Key words: grain legumes, mutagenesis, nitrogen fixation, plant breeding

Abstract

Genetic variation among existing cultivars and in germplasm collections is the outcome of selection during evolution and plant breeding. Mutagenesis offers the plant breeder a chance to tackle unconventional objectives, particularly those that were at a selection disadvantage in the past. Effective mutagens are available, but the bottleneck is the effective selection of rare desired variants from large mutagenized populations. Selection methods must be non-destructive. Grain legume mutation breeding has already led to improved cultivars with higher yield, better grain quality, or stronger resistance to pathogens. Many mutations affecting nitrogen fixation related traits have also been reported. Some could be useful in breeding better cultivars, but the majority are being used to study the factors interacting in the complex process of symbiotic nitrogen fixation and to improve the strategy for producing cultivars with better fixation capacity.

Introduction

The Plant Breeding and Genetics Section of the Joint FAO/IAEA Division, since its establishment in 1964, tries to help plant breeders to develop improved cultivars through the use of induced mutations. The work has not been restricted to any plant species and grain legumes have always had a prominent place (Micke and Swiecicki, 1988 Micke et al., 1990). Mutation induction supplies breeders with genetic variants from which useful types may be selected. Thus, 'mutation breeding' requires an effective mutagen treatment and an efficient way of selection. The genetic variation from mutagenesis differs from that existing in germplasm collections or obtainable from crossing in that it is not yet selected by nature or man and therefore contains also traits which were not favoured during evolution or past plant breeding. This offers the breeder a chance to enhance unconventional breeding objectives. There are also good prospects for improving adaptation of recently introduced crops to unsuitable environments, as well as for the 'domestication' of promising wild species for agricultural or horticultural uses (FAO/IAEA, 1986). Mutation breeding therefore appears to be an appropriate approach not only for genetic improvement of grain legumes in general but also for improving their symbiotic nitrogen fixation (Micke 1984, 1988).

The problem

Many of the leguminous species cultivated for grain production are only partly domesticated and not well suited for monoculture (e.g. poor lodging resistance). They possess the ability to fix nitrogen in symbiotic association with bacteria – a considerable advantage under the inter-species competition prevailing during natural evolution – but the amount fixed under nitrogen deficient conditions often satisfies only the quantity needed for survival and offspring production, and seldom is sufficient for a good yield (Bliss and Miller, 1988). Where sufficient soil nitrogen is available, the competitive advantage of nitrogen fixing plants disappears. This seems to be an example of good evolutionary adaptation that under high levels of applied nitrogen, N_2 fixation is progressively inhibited.

As a consequence, in developing countries where farmers cannot easily afford nitrogen fertilizers, grain legume cultivation has moved to soils of marginal fertility, leaving the better ones to the cereals and other nitrogen responsive crops. Increased cereal grain production was a desirable result of the so-called 'Green Revolution', but its undesirable side effect was a decrease in the production and steep increases in the price of pulses, which traditionally provide a major source of protein in the diet in many developing countries. To reverse this unfortunate trend (Borlaug, 1972), it is necessary to increase yields of grain legumes through breeding, possibly also through more effective symbiotic nitrogen fixation and through eliminating the inhibition of nitrogen fixation occurring when soils are rich in nitrogen or when large amounts of N-fertilizers are applied. A grain harvest of $4\,t\,ha^{-1}$ – which could make grain legumes competitive to high yielding varieties of cereals – would remove about $150–200\,kg\,N\,ha^{-1}$ from the field. Since for most grain legumes only about half of the absorbed nitrogen is recovered in the grain, there should be an amount of $300–400\,kg\,N\,ha^{-1}$ available to the plants. Such amounts have been found to be fixed by some species under favourable circumstances, therefore the objective is not unrealistic. (Coale et al., 1985; Evans et al., 1987; Herridge and Doyle, 1988; Jensen, 1987; Phillips, 1980; Sinha et al., 1983).

Practice of mutation breeding

Mutation induction raises the natural mutation rates 10–100 fold, so that the breeder has an improved opportunity to observe a reasonable number of mutants in a limited nursery. Even then, populations of 100,000 plants (or a multiple of this) must be screened for the desired variants because mutation rates per locus/per generation are estimated to be around 1×10^{-4} to 1×10^{-5}. Thus, as ample potent mutagens are available, selection of desirable mutants has become the bottleneck for mutation breeding. This is not so critical when the desired trait is easy to see as e.g. early flowering or altered plant architecture, but selection procedures limit the prospects of mutation breeding when more sophisticated tests are required as for assessing nitrogen fixation of numerous individual plants or single plant progenies.

In the practice of mutation breeding, one should be aware of complex genetic and physiological damages caused initially by any mutagen treatment. The first generation (M_1) developed from treated seeds, e.g. suffers from growth inhibition, is partly sterile, and may lose many plants before flowering and seed set. Moreover, these M_1 plants are genetic chimeras and may carry a mutation only in a fairly small sector. The interaction with micro-organisms is not normal and therefore one has to protect M_1 plants from pathogens and assure optimal nutrition. Selection for mutants should start only in the second generation and concern single M_2 plants and their progenies in the M_3 generation. The selection procedure must be non-destructive in the sense that viable seeds are obtained from selected plants.

Achievements of legume mutation breeding

Many objectives have been reached by mutation breeding of grain legumes. Many mutant varieties of common bean have been released (Table 1) as well as soybean (43), groundnut (33), pea (27), lupin (18), mungbean (9), cowpea (9), faba bean (8), chickpea (7), pigeon pea (5), etc., which possess many different improved characters such as disease and insect pest resistance, earlier or later flowering, higher yield, higher protein content or less toxic compounds. (FAO/IAEA, 1988; Jaranowski and Micke, 1985; Micke et al., 1985; Micke and Swiecicki, 1988).

Prospects of mutation breeding for better nitrogen fixation

Few of the varieties released so far, are claimed to possess higher nitrogen fixation (Ladha et al., 1983; Schroeder et al., 1988). One should, how-

Table 1. Improved cultivars of common bean developed by mutation breeding

Species and name of cultivar	Country and year of release	Mutagen used	Improved characters
Phaseolus vulgaris L. (common bean)			
Universal	Germany, 1950	x-rays	early maturing, higher yield, disease resistant
Sanilac	USA, 1956	x-rays	bush type
Unima	Germany, 1957	x-rays & cross	disease resistant
Seaway	USA, 1960	x-rays & cross	bush type, short duration, virus resistant
Gratiot	USA, 1962	x-rays & cross	bush type, stiffer stem, higher protein content
Saparke 75	USSR, 1967	gamma rays	higher yield, better disease resistance
Seafarer	USA, 1967	x-rays & cross	very early maturing bush type, disease resistant
Pusa Parvati	India, 1970	x-rays	early, bush type, higher yield
Alfa	CSFR, 1972	EMS	white seed colour, higher yield and protein content
Giza 80	Egypt, 1980	gamma rays	higher yield, larger grain, rust resistant
Muhranula	USSR, 1982	EI	earlier
Ouray	USA, 1982	x-rays & cross	bush type, better disease resistance
Markovskaya 8	USSR, 1985	gamma rays	white seeds, earlier
Mogano	Italy, 1985	EMS	uniform seed colour, dwarf type, high yield
Montalbano	Italy, 1985	EMS	uniform seed colour, high yield
CAP 1070	Brazil, 1986	gamma rays	bush type, earlier maturity
Mitchell	Canada, 1986	x-rays & cross	higher yield
Neptune	USA, 1986	x-rays & cross	bush type

ever, assume that the good vigour and high yield reported so often could only be obtained under conditions of limited soil nitrogen, if a mutant variety possesses a satisfactory nitrogen fixation capacity. In most cases, the nitrogen fixation capacity has not yet been examined. There are, however, a number of reports claiming improved nodulation and/or N_2 fixation of induced mutants not yet released as commercial cultivars, e.g. in mungbean (Rosaiah et al., 1987; Thakare et al., 1983), cowpea (Mahna et al., 1990), blackgram (Mahna et al., 1990; Sing and Raghuvanshi, 1987), faba bean (Brunner and Zapata, 1984).

A rather frequently observed type of spontaneous as well as induced mutation removes the regulation of nodule density and leads to 'supernodulation', e.g. in soybean (Carroll et al., 1985), pea (Jacobsen, 1984; Rosendahl et al., 1989) common bean (Park and Buttery, 1988). Also rather numerous are the reports about mutants resistant to nodulation (non-nod). Such mutants are easily detectable by their yellowish appearance and were observed, e.g. in soybean (Carroll et al., 1986; Mathews et al., 1989), pea (Engvild, 1987; Kneen and La Rue, 1984), peanut (Gorbet and Burton, 1979), chickpea (Davis et al., 1985), bean (Davis et al., 1988).

Mutants with either more or fewer nodules may provide some more insight into the complex interactions between host plants and bacteria (Postma et al., 1988). Such mutants should also be good experimental materials for physiological studies, including environmental interactions. Most of the 'supernodulating' mutants reported have some tolerance to soil nitrate, but so far none has shown a substantially higher nitrogen fixation. Probably, the 'super-nod' mutants cannot be compared with the original type for judging specifically the effect of more nodules on nitrogen fixation, since other traits besides those concerning specifically the nodulation have also been changed. Therefore, near-isogenic lines should be developed through backcrossing to produce materials suitable for studying only the effect of abundant nodules. It appears that host plants have effective compensation mechanisms to maintain a balanced C/N metabolism. Other factors controlling 'nitrogen fixation supportive traits' may have to be changed as well to make use of the potential of more nodule formation. Selection for enhanced nitrogen fixation may be direct or indirect (Bliss and Miller, 1988). Particularly in mutation breeding, where single plants have to be assessed among large populations, the selection method must be non-destructive. Analysing xylem sap composition is effective in some species (Herridge, 1988; Pate et al., 1988; Peoples et al., 1987; Rosendahl and Jakobsen, 1987) and fluorescence measurements may be another possibility (Paul and Planchon, 1990).

Many heritable morphological and physiological traits will indirectly affect nitrogen fixation. This must be kept in mind not only when comparing different varieties but also when developing a breeding strategy for improving fixation. If the breeder desires a plant with an altered plant architecture (e.g. determinate type of growth) and an increased 'harvest index' (less leaves and stems, more pods and grains), the nitrogen fixation should be monitored in an effort not to lose the previous levels. In this context, it might be appropriate to consider the amount of nitrogen fixed per m^2 of cropped land as more important than the amount per plant. There appears to persist a controversy on the relevance of selecting plants for 'higher percentage N derived from fixation'. It was found, that 'increased amount of N_2 fixed' is closely related to the biomass, plant architecture and crop duration while the 'N percentage derived from fixation' is depending more often on soil N than on the host genotype (Beringer, 1988; Betts and Herridge, 1987; Coale et al., 1985). This matter needs clarification to allow an optimal plant breeding strategy to be developed.

As the amount of nitrogen recovered in the seeds on the average is only 50% of the total N of the plant, it seems worth considering a selection for better nitrogen translocation from vegetative plant parts to the seed (Herridge and Doyle, 1988; Jensen, 1987; Sinha et al., 1983). Unfortunately, nitrogen translocation leads to senescence and early senescence reduces nitrogen fixation (Devries et al., 1989). Late senescence on the other hand increases crop duration and may affect negatively the harvestability of the grain.

In most farming systems, legumes are part of the crop rotation but in some systems they are also used as intercrops. Unfortunately, nitrogen fixation appears to have evolved as a 'selfish' process where companion plants derive little benefit from the nitrogen fixed by legumes. In assessing the nitrogen balance of cereal-legume rotations, it was found that in some circumstances grain legumes may leave even less nitrogen to the subsequent crop than that which they have found as residual nitrogen from the preceding crop (Bezdicek et al., 1986; Eaglesham et al., 1982). Thus, efforts of breeders in co-operation with plant physiologists should also be directed towards legume varieties that are more efficient in providing nitrogen to companion crops. The same varieties should not show an inhibition of nitrogen fixation when nitrogen fertilizers are applied (Herridge and Betts, 1988). These desired characteristics would be contrary to the principles of competition under natural evolution and therefore may not be found in 'natural' germplasm. Thus, mutagenesis would seem to be the best approach to obtain such desired genetic variation.

References

Beringer J E, Bisseling T A and La Rue T A 1988 Improving symbiotic nitrogen fixation through the genetic manipulation of *Rhizobium* and legume host plants. *In* World Crops: Cool Season Food Legumes. Ed. R J Summerfield. pp 691–702. Kluwer Academic Publishers, Dordrecht.

Betts J H and Herridge D F 1987 Isolation of soybean lines capable of nodulation and nitrogen fixation under high levels of nitrate supply. Crop Sci. 27, 1156–1161.

Bezdicek D F, Root C S, Smith S C and Muehlbauer F J 1986 The influence of residual soil and mineralized-nitrogen on legume N_2 fixation and contribution to the soil nitrogen pool. Poster presented at the International Food Legume Research Conference Spokane, WA, USA, 7–11 July 1986.

Bliss F A and Miller J C Jr 1988 Selecting and breeding grain legumes for enhanced nitrogen fixation. *In* World Crops: Cool Season Food Legumes. Ed. R J Summerfield. pp 1001–1012. Kluwer Academic Publishers, Dordrecht.

Borlaug N E 1972 Building a protein revolution on grain legumes. *In* Nutritional Improvement of Food Legumes by Breeding. Protein Advisory Group of the UN System, New York. Proceedings of a symposium Rome 1972. pp 7–11.

Brunner H and Zapata F 1984 Quantitative assessment of symbiotic nitrogen fixation in diverse mutant lines of field bean (*Vicia faba minor*). Plant and Soil 82, 407–413.

Carroll B J, Mc Neil D L and Gresshoff P M 1985 A supernodulation and nitrate tolerant symbiotic (*nts*) soybean mutant. Plant Physiol. 78, 34–40.

Carroll B J, Mc Neil D L and Gresshoff P M 1986 Mutagenesis of soybean (*Glycine max* (L.) Merr.) and the isolation of non-nodulating mutants. Plant Sci. 47, 109–114.

Coale F J, Meisinger J J and Wiebold W J 1985 Effects of plant breeding and selection on yields and nitrogen fixation in soybeans under two soil nitrogen regimes. Plant and Soil 86, 357–367.

Davis J H C, Giller K E, Kipe-Nolt J A and Awah M 1988

Non-nodulating mutants in common bean. Crop Sci. 28, 859–860.

Davis T M, Foster K W and Phillips D A 1985 Nodulation mutants in chickpea. Crop Sci. 25, 345–348.

Devries J D, Bennett J M, Boote K J, Albrecht S L and Maliro C E 1989 Nitrogen accumulation and partitioning by three grain legumes in response to soil water deficits. Field Crops Res. 22, 33–44.

Eaglesham A R J, Ayanaba A, Ranga Rao V and Eskew D L 1982 Mineral N effects on cowpea and soybean crops in a Nigerian soil. II. Amounts of N fixed and accrual to the soil. Plant and Soil 68, 183–192.

Engvild K C 1987 Nodulation and nitrogen fixation mutants of pea, *Pisum sativum*. Theor. Appl. Genet. 74, 711–713.

Evans J, Turner G L, O'Connor G E and Bergersen F J 1987 Nitrogen fixation and accretion of soil nitrogen by field-grown lupins (*Lupinus angustifolius*). Field Crops Res. 16, 309–322.

FAO/IAEA 1986 Plant Domestication by Induced Mutations. IAEA, Vienna.

FAO/IAEA 1988 Improvement of Grain Legume Production Using Induced Mutations. IAEA, Vienna.

Gorbet D W and Burton J C 1979 A non-nodulating peanut. Crop Sci. 19, 727–728.

Herridge D F 1988 The narrow-leafed lupin (*Lupinus angustifolius* L.) as a nitrogen-fixing rotation crop for cereal production. I. Indices of nitrogen fixation. Aust. J. Agric. Res. 39, 1003–1015.

Herridge D F and Betts J H 1988 Field evaluation of soybean genotypes selected for enhanced capacity to nodulate and fix nitrogen in the presence of nitrate. Plant and Soil 110, 129–135.

Herridge D F and Doyle A D 1988 The narrow-leafed lupin (*Lupinus angustifolius* L.) as a nitrogen-fixing rotation crop for cereal production. II. Estimates of fixation by field-grown crops. Aust. J. Agric. Res. 39, 1017–1028.

Jacobsen E 1984 Modification of symbiotic interaction of pea (*Pisum sativum* L.) and *Rhizobium leguminosarum* by induced mutations. Plant and Soil 82, 427–438.

Jaranowski J and Micke A 1985 Mutation Breeding in Peas. Mutation Breeding Review No. 2. IAEA, Vienna. 23 p.

Jensen E S 1987 Seasonal patterns of growth and nitrogen fixation in field grown pea. Plant and Soil 101, 29–37.

Kneen B E and La Rue T A 1984 Nodulation resistant mutant of *Pisum sativum* (L.). J. Hered. 75, 238–240.

Ladha M L, Johari R P, Sharma N D and Mehta S L 1983 Nitrogen fixation in relation to dark CO_2 fixation in developing nodules of *Arachis hypogaea* L. mutants. Ind. J. Exp. Biol. 21, 629–632.

Mahna S K, Bhargava L and Mohan L 1990 Alkaline azide mutagenicity in cowpea. Mut. Breed. Newsl. 36, 6–7.

Mahna S K, Garg R and Parvateesam M 1990 Nodulation studies with induced mutants of blackgram (*Vigna mungo* L.). Mut. Breed. Newsl. 36, 9–10.

Mathews A, Carroll B J and Gresshoff P M 1989 A new non-nodulation gene in soybean. J. Hered. 80, 357–360.

Micke A 1984 Mutation breeding of grain legumes. Plant and Soil 82, 337–357.

Micke A 1988 Genetic improvement of food legumes in developing countries by mutation induction. *In* World Crops: Cool Season Food Legumes. Ed. R J Summerfield. pp 1021–1047. Kluwer Academic Publishers, Dordrecht.

Micke A, Donini B and Maluszynski M 1985 Plant cultivars derived from mutation induction or the use of induced mutants in cross breeding. Mutation Breeding Review No. 3. IAEA, Vienna. 92 p.

Micke A, Donini B and Maluszynski M 1990 Induced Mutations for Crop Improvement. Mutation Breeding Review No. 7. IAEA, Vienna. 41 p.

Micke A and Swiecicki W 1988 Induced mutations in lupins. *In* Proceedings of 5th Intern. Lupin Conference, Poznan, Poland. pp 110–127.

Park S J and Buttery B R 1988 Nodulation mutants of white bean (*Phaseolus vulgaris* L.) induced by ethyl-methane sulfonate. Can. J. Plant Sci. 68, 199–202.

Pate J S, Atkins C A, Peoples M B and Herridge D F 1988 Partitioning of carbon and nitrogen in the nodulated grain legume: Principles, processes and regulation. *In* World Crops: Cool Season Food Legumes. Ed. R J Summerfield. pp 751–765. Kluwer Academic Publishers, Dordrecht.

Paul M H and Planchon C 1990 Chlorophyll fluorescence as a tool in soybean improvement for N_2 fixation efficiency. Euphytica 45, 43–47.

Peoples M B, Sudin M N and Herridge D F 1987 Translocation of nitrogenous compounds in symbiotic and nitrate-fed amide-exporting legumes. J. Exp. Bot. 38, 567–579.

Phillips D A 1980 Efficiency of symbiotic nitrogen fixation in legumes. Annu. Rev. Plant Physiol. 31, 29–49.

Postma J G, Jacobsen E and Feenstra W J 1988 Three pea mutants with an altered nodulation studied by genetic analysis and grafting. J. Plant Physiol. 132, 424–430.

Rosaiah G, Kumari D S, Satyanarayana A and Seenaiah P 1987 An improvement in nitrogen-fixing ability of green gram through gamma irradiation. Ind. J. Agric. Sci. 57, 271–273.

Rosendahl L and Jakobsen I 1987 *Rhizobium* strain effects on pea: The relation between nitrogen accumulation, phosphoenolpyruvate carboxylase activity in nodules and asparagine in root bleeding sap. Physiol. Plant. 71, 281–286.

Rosendahl L, Vance C P, Miller S S and Jacobsen E 1989 Nodule physiology of a supernodulating pea mutant. Physiol. Plant. 77, 606–612.

Schroeder H E, Gibson A H, Oram R N and Shaikh M A Q 1988 Seed protein and nitrogen fixation in chickpea mutant variety 'Hyprosola'. J. Sci. Food Agric. 44, 31–41.

Sing P K and Raghuvanshi S S 1987 A higher yielding mutant of black gram with improved nodule formation. Mut. Breed. Newsl. 30, 16–17.

Sinha S K, Lal B, Koundal K R and Chopra R K 1983 Nitrogen fixation does not limit yield in pulses. Current Sci. 52, 916–918.

Thakare R G, David K A V, Thomas J and Bhatia C R 1983 Increased nitrogen fixation in an induced mungbean mutant. Mut. Breed. Newsl. 22, 7–8.

Plant and Soil **152**: 87–91, 1993.

PLSO BNF08

Yield and biological nitrogen fixation of common bean (*Phaseolus vulgaris* L.) in Peru

A. MANRIQUE, K. MANRIQUE and J. NAKAHODO
Universidad Nacional Agraria La Molina, Apartado 456, Lima, Peru

Key words: common bean, ^{15}N isotope dilution, nitrogen fixation, *Phaseolus vulgaris* L.

Abstract

Two field experiments were performed to evaluate the nitrogen fixation potential of twenty common bean cultivars and breeding lines during summer and winter seasons of 1986 and 1988, respectively. The ^{15}N isotope dilution method was used to quantify N_2 fixation. The cultivars and breeding lines were variable in terms of their N_2 fixation. The cv. Caballero was very efficient, with more than 50% N derived from the atmosphere and 60–80 kg N ha^{-1} fixed in both seasons. Other cultivars were less efficient, since the poorest ones derived less than 30% of their nitrogen from the atmosphere and fixed less than 20 kg N ha^{-1}. After additional testing the best cultivars may be used directly by the farmers for cultivation. The experiments have provided information about which genotypes may be used to breed for enhanced fixation in common bean.

Introduction

Beans have been used as human food for more than 7000 years in Peru according to archeologists using the ^{14}C technique to measure the age of beans found in a tomb of the Callejon de Huaylas in the Ancash Highlands (Voysest, 1983). Along with corn, beans were one of the most important sources of protein for the ancient cultures of Peru, i.e., Mochicas, Chavin, Wari and Paracas, even before the Incas culture. Currently, more than 60,000 ha are cultivated to common bean in Peru, of which approximately 90% are harvested as dry seeds (Table 1).

Common beans are grown mainly under irrigation in the coastal area of Peru either in rotation with other crops or as a sole crop. In the highlands and in the tropical forest they are grown under rainfed conditions. In general, yields are very low, from 600 to 2,000 kg ha^{-1}, which is due mainly to diseases or lack of improved varieties.

Two programmes organized by the Ministry of Agriculture (INIAA) and the National Agriculture University at La Molina (UNALM) have been conducted to improve common bean production. The main objectives are to increase yields through greater availability of seed with resistance to diseases, insects, drought and with improved biological nitrogen fixation. The objective of the experiments was to investigate the nitrogen fixing ability of cultivars and breeding lines of common bean to be recommended to farmers for production or to be used in breeding programmes.

Materials and methods

The experiments performed in this study were conducted in the field of the National Agricultural University at La Molina during the winter (April 24–August 28, 1986) and summer (December 21–March 17, 1988) seasons. During the above years twenty and twenty-two bean lines, respectively, including 11 commercial cultivars and 11 breeding lines, which were induced mutants from the Canario and Bayo cultivars were

Table 1. Area, production and yield of leguminous crops in Peru

Leguminous crops		No. of hectares	Total yield (t)	Yield (kg/ha)
Bean (*Phaseolus vulgaris* L.)	Green:	5 924	8 565	1 446
	Grain:	57 014	47 693	836
	Total:	62 938		
Pea (*Pisum sativum* L.)	Green:	10 643	30 342	2 851
	Grain:	23 681	18 149	766
	Total:	34 324		
Chiclayo bean (cowpea sp.)	Green:	250	250	1 000
	Grain:	2 220	1 568	
	Total:	2 470		
Palo bean (*Cajanus cajan* L.)	Green:	205	344	1 676
	Grain:	329	358	1 088
	Total:	534		
Zarandaja (*Dolichus lablab* L.)	Green:	470	802	1 705
	Grain:	3 205	2 698	841
	Total:	3 675		
Lima bean (*Phaseolus lunatus* L.)	Green:	1 667	6 270	3 761
	Grain:	4 342	3 624	834
	Total:	6 019		
Broad bean (*Vicia faba* L.)	Green:	7 325	33 095	4 518
	Grain:	22 827	20 628	904
	Total:	30 152		
Chocho-tarhui (*Lupinus mutabilis*)	Grain:	4 500	5 246	1 166
Chick-pea (*Cicer arietinum* L.)	Grain:	2 671	1 937	725
Lens (*Lens esculentum* L.)	Grain:	2 059	1 229	595
Lautao (*Vigna radiata* L.)	Grain:	902	797	884
Nuña (*Phaseolus vulgaris* L.)	Grain:	427	265	620

tested (Table 2). Canario is one of the two most important bean cultivars in Peru, being cultivated on 40% of the coastal area. It is a type I cultivar of medium precocity having large, oval, yellow seeds. The second cultivar, Bayo is also cultivated on approximately 40% of the coastal area. It is a type III cultivar of medium precocity having large, oval, brown seeds. Caraota Brasileño and Caraota Venezolano are type II cultivars of medium precocity and black seeds. These are less important cultivars grown mainly in the summer season in the coastal and tropical areas. Blanco, Panamito, Panamito Grande and Caballero are all cultivars with white seeds, and are cultivated mainly in the coastal area during the winter season.

The mutant lines of Canario and Bayo were identified during a previous FAO/IAEA programme on "Improvement of Beans through Induced Mutations" with the main objective being to change the growth habits of these cultivars. Seven lines were selected from Canario Divex and four from Bayo after seven cycles of selection (Manrique and Nakahodo, 1988). Four non-fixing crops i.e. wheat, barley, sorghum and non-nodulating soybean were used as references for measuring N_2 fixation.

After planting, a liquid mixture of five local

Table 2. Common bean (*Phaseolus vulgaris* L.) cultivars or lines tested in the present study

1. 11 Commercial varieties of beans

   Caraota Brasileño
   Caraota Venezolano
   Negro
   Blanco
   Panamito Grande
   Panamito
   Caballero
   Canario Divex
   Red Cloud
   Bayo Normal
   Tortola Diana

2. 7 Selected lines of Canario Divex

   CG. 58-13
   CG. 62-4
   CG. 62-13
   CG. 187-2
   CG. 25-8
   CG. 62-2-6
   CG. 252-17

3. 4 Selected lines of Bayo

   Bayo G 10.0-5
   Bayo G 7.5-2
   Bayo G 5.0-1 (erecto)
   Bayo G 7.5-9 (erecto)

strains of *Rhizobium leguminosarum* biovar. *phaseoli* was applied to the soil as an inoculant for all cultivars. The experiment was set up as a randomized block design with each treatment replicated six times. Each of the bean lines was planted in 1.5 m rows with 10 cm spacing between plants and 40 cm between rows. After planting, 10 kg N ha^{-1} solution of 10.075% ^{15}N atom excess ammonium sulphate was applied to the plots. The field was maintained clean of weeds and irrigated when necessary. The plots were harvested at physiological maturity of the cultivars with the shortest maturity time. They were separated into pods and straw; then dry matter production was determined. The samples were milled and analysed for %N by the Kjeldahl procedure (Eastin, 1978) and %^{15}N atom excesses by mass-spectrometry (Fiedler and Proksch, 1975) at the IAEA laboratory in Seibersdorf, Austria. Nitrogen fixation was calculated from the %^{15}N atom excess data using the ^{15}N isotope dilution method (Fried and Middelboe, 1977; Hardarson, 1990; McAuliffe et al., 1958).

Results and discussion

Dry matter yield, %N, N yield, %^{15}N atom excesses, % Ndfa and total Ndfa measured in the two seasons are presented in Tables 3 and 4. Nitrogen fixation measured by the ^{15}N isotope dilution method with wheat as a non-fixing reference crop, ranged from 24 to 56% Ndfa and 15 to 59 kg N ha^{-1} in the winter season, and from 13 to 56% Ndfa and from 7 to 81 kg N ha^{-1} during the summer season. Only the amount of N_2 fixed was statistically significant ($p < 0.05$) in both seasons. Caballero was the best cultivar for N_2 fixation in both seasons, with Caraota Brasilano and Blanco performing well in the winter season and some of the Bayo and Canario lines fixing well during the summer season. There were differences in the rankings of cultivars and lines observed in the two seasons.

In the winter of 1986, dry matter yield production was significantly correlated ($p < 0.01$) to the amount of N_2 fixed, but no correlation was found between dry matter yield and % Ndfa (Table 5). Similar results were observed in the summer of 1988. Nitrogen yield was highly correlated to the amount of fixed N_2 than % Ndfa in both seasons. In both experiments highly significant ($p < 0.001$) correlations were measured between percentage and amount of N_2 fixed.

The differences in rankings of bean cultivars and lines between seasons were observed mainly for dry matter yield, %^{15}N a.e., and % Ndfa. Caballero fixed well in both seasons and other cultivars such as Panamito Grande, Caraota Venezolano, Red Kloud and Panamito showed similar ranking in both seasons. Some cultivars e.g., Caraota Brasilano, Blanco, Negro, and Canario Divex and many of the Canario breeding lines, had very different ranking in the two seasons (Tables 3 and 4). Correlations between N yield and amount of fixed N_2 ($p < 0.001$) were significant (Table 5), but not those between dry matter yield, %^{15}N a.e., and % Ndfa in the two seasons.

Table 3. Yield and nitrogen fixation of twenty cultivars or lines of common bean grown in the winter season of 1986 at La Molina

Entry	Cultivar	Dry matter ($t\ ha^{-1}$)	%N	N yield ($kg\ ha^{-1}$)	% 15N a.e.	% Ndfa	Fixed N_2 ($kg\ ha^{-1}$)
1	Caballero	5.1	2.1	107	0.050	56	59
2	Caraota	4.3	2.2	95	0.050	56	53
3	Blanco	3.4	2.2	75	0.055	51	38
4	Canario G-58-13 (14)	4.0	2.0	80	0.061	46	37
5	Canario Divex	3.3	2.1	69	0.062	45	31
6	Canario G-62-13 (32)	3.2	2.0	64	0.062	45	28
7	Canario G-62-4 (27)	3.9	2.0	78	0.064	43	34
8	Panamito Grande	3.7	2.1	78	0.064	43	34
9	Canario G-187-2 (35)	3.9	2.0	78	0.065	43	34
10	Bayo G-7.5-2	4.0	2.1	84	0.068	40	34
11	Bayo G-7.5-9e	4.6	1.8	83	0.069	39	33
12	Canario G-25-18 (4)	3.7	2.0	74	0.071	38	28
13	Negro	3.5	1.9	66	0.075	34	22
14	Caraota Venezolano	3.1	1.7	53	0.075	34	18
15	Panamito	2.9	2.0	58	0.077	32	19
16	Red Kloud	2.7	2.3	62	0.078	32	19
17	Canario G-62-2-6 (26)	3.6	1.9	68	0.078	32	22
18	Bayo G-10.G-5	4.1	2.0	82	0.080	30	24
19	Bayo G-5.0.1e	4.7	1.9	89	0.082	28	25
20	Canario G-252-17 (37)	3.4	1.9	65	0.087	24	15
	Wheat	2.2	0.8	18	0.114		
LSD	0.05					NS	25

Table 4. Yield and nitrogen fixation of twenty two cultivars or lines of common bean grown in the summer season of 1988 at La Molina

Entry	Cultivar	Dry matter ($t\ ha^{-1}$)	%N	N yield ($kg\ ha^{-1}$)	% 15N a.e.	% Ndfa	Fixed N_2 ($kg\ ha^{-1}$)	Grain yield ($kg\ ha^{-1}$)
1	Caballero	5.6	2.5	146	0.139	56	81	1,561
2	Bayo Normal	4.3	2.7	116	0.144	54	63	817
3	Bayo G-10.0-5	3.0	2.9	87	0.159	49	43	801
4	Bayo G-7.5-9	2.9	2.5	73	0.164	48	35	979
5	Canario G-62-2-6 (26)	2.9	2.6	73	0.174	45	32	1,043
6	Canario G-62-4 (27)	2.9	2.6	75	0.179	43	32	983
7	Negro	2.7	2.1	57	0.185	41	27	1,225
8	Caraota Venezolano	2.7	2.1	57	0.186	40	23	1,002
9	Canario G-252-17 (37)	2.6	2.6	68	0.186	40	27	1,354
10	Canario Divex	2.6	2.7	70	0.186	40	28	1,265
11	Canario G-62-13 (32)	2.9	2.5	73	0.189	40	29	1,359
12	Tortola Diana	3.9	2.2	86	0.192	39	33	1,733
13	Panamito	3.1	1.9	59	0.197	37	22	658
14	Canario G-25-18 (4)	2.7	2.6	70	0.200	36	25	867
15	Canario G-58-13 (14)	2.8	2.7	76	0.201	36	27	1,007
16	Blanco	2.7	2.2	59	0.206	34	20	649
17	Bayo G-7.5-2	2.9	2.6	75	0.206	34	25	1,574
18	Canario G-187-2 (35)	3.2	2.5	80	0.210	33	26	1,478
19	Panamito Grande	2.7	2.3	62	0.216	31	19	1,129
20	Caraota Brasileño	2.1	2.5	53	0.229	27	14	1,433
21	Bayo G-5.0-1e	2.0	2.8	56	0.234	25	14	780
22	Red Kloud	2.4	2.2	53	0.272	13	7	1,250
	Wheat	0.4	1.8	7	0.314			1,134

Table 5. Correlations of various measured parameters in the winter of 1986 and summer of 1988

	1986		1988				
	% Ndfa	Fixed N_2	Dry matter	N yield	% ^{15}N atom excess	% Ndfa	Fixed N_2
1986							
Dry matter	0.24 NS	0.58**	0.40 NS				
N yield	0.44*	0.72***		0.58**			
% ^{15}N atom excess					0.10 NS		
% Ndfa		0.85**				0.17 NS	
Fixed N_2							0.59**
1988							
Dry matter						0.42 NS	0.84**
N yield						0.43 NS	0.85**
% Ndfa							0.73**

NS: Non-significant
*, **, ***: Significant at $p < 0.05$, $p < 0.01$ and $p < 0.001$ levels, respectively.

Although common bean often is considered to be a poor fixer of atmospheric N_2, substantial variation among cultivars and induced mutant lines derived from them was observed during two growing seasons. In each experiment, the highest amount of total Ndfa was about 60 kg ha^{-1}, which represents about one-half of the total plant N as being derived from the atmosphere. It was not surprising that grain yield was not highly positively correlated with parameters of N_2 fixation, since the plants ranged from type I to type III, and all were not equally well adapted. The range of variability for important traits such as yield and N_2 fixation parameters among the different plant types and commercial classes offer the possibility of increasing N_2 fixation in improved cultivars above that found in the standard cultivars. Natural variation and induced mutants, provided that other traits are not changed deleteriously, can be used for hybridization and selection experiments to produce improved cultivars.

References

Eastin E F 1978 Total nitrogen determination for plant material, Anal. Biochem. 85, 591–594.

Fiedler R and Proksch G 1975 The determination of ^{15}N by emission and mass spectrometry in biochemical analysis: A review. Anal. Chim. Acta 78, 1–62.

Fried M and Middelboe V 1977 Measurement of the amount of nitrogen fixed by a legume crop. Plant and Soil 47, 713–715.

Hardarson G 1990 Use of Nuclear Techniques in Studies of Soil-Plant Relationships. Training Course Series No. 2. IAEA, Vienna.

Manrique A, Nakahodo J and Manrique K 1988 Improvement of frijol Bayo (*Phaseolus vulgaris* L.) through induced mutations *In* Improvement of Grain Legume Production using Induced Mutations. pp 157–166. IAEA, Vienna.

McAuliffe C, Chamblee D S, Uribe-Arango H and Woodhouse W W 1958 Influence of inorganic nitrogen on nitrogen fixation by legumes as revealed by ^{15}N. Agron. J. 50, 334–337.

Voysest O 1983 Variedades de frijol en America Latina y su origen. Centro Internacional de Agricultura Tropical. Cali, Colombia. 87 p.

Plant and Soil **152**: 93–101, 1993.

PLSO BNF09

Variability in nitrogen fixation of common bean (*Phaseolus vulgaris* L.) intercropped with maize

S.M. TSAI, P.M. DA SILVA, W.L. CABEZAS and R. BONETTI
Centro de Energia Nuclear na Agricultura, University of São Paulo, P.O. Box 96, Piracicaba, CEP 13400-970, São Paulo, Brazil

Key words: common bean, maize, intercropping, nitrogen fixation, ^{15}N isotope, ^{15}N natural variation, *Rhizobium*

Abstract

Thirty one selected bean lines were evaluated in the field for ability to support N_2 fixation when intercropped with maize which received 0, 30 and 60 kg N ha^{-1} as ammonium sulphate. The amount of fixed N_2 was estimated using the natural variation of ^{15}N and wheat as the standard non-fixing crop. Nitrogen as low as 15 kg N ha^{-1} at sowing suppressed nodule weight and activity (acetylene reduction activity) but not nodule number, suggesting that the main effect of mineral N was on nodule development and function. ^{15}N data revealed a high potential of the bean genotypes to fix N_2, with the most promising ones averaging between 50–60% of seed N coming from fixation. Bean lines CNF-480, Puebla-152, Mexico-309, Negro Argel, CNF-178, Venezuela-350 and WBR22-3, WBR22-50 and WBR22-55 were ranked as good fixers.

Introduction

Nitrogen fixation in common bean (*Phaseolus vulgaris* L.) has been quantified by several methods, including the ^{15}N isotope technique. Values ranging from 20 to 124 kg N ha^{-1} were obtained in field-grown beans at contrasting environments (Graham and Halliday, 1977; Rennie and Kemp, 1983a,b; Saito, 1982; Westermann and Kolar, 1978). Most commonly-used bean cultivars exhibit a high dependence on N fertilizers for growth and yield, and show considerable variation in their ability to nodulate and fix N_2, with the %N derived from atmosphere (%Ndfa) ranging from 68–72% for the superior cultivars (Rennie and Kemp, 1983a,b; Ruschel et al., 1982; Westermann and Kolar, 1978). The average of N_2 fixation by field-grown beans in Brazilian tropical soils is less than 60 kg N ha^{-1} (Duque, et al., 1985; Saito, 1982), with only 30% of the fixed N_2 converted to seed N (Ruschel et al., 1982).

Most bean improvement programs in Latin America have developed cultivars for high yield potential, using fertile, N-rich soils. This strategy, however, may suppress expression of desirable traits for N_2 fixation in the plant. Inoculation of common bean with effective *Rhizobium* strains has been shown to result in increased plant weight, nitrogen level in the plant and yield (Saito, 1982). However, under field conditions the *Rhizobium*-symbiosis often has been less efficient due to various limiting factors (Franco and Döbereiner, 1967; Graham, 1981). Graham and Temple (1984) reviewed the status of N_2 fixation in beans and emphasized the need to assess nursery materials under conditions which permit nodulation and N_2 fixation with the native soil rhizobia rather than dependence on N fertilizers.

The high costs of N fertilizers and in contrast, the prevalent low-N input cropping systems of bean in Latin America raise the need for enhanced levels of N_2 fixation to fulfill N requirements of high yielding cultivars (Pereira et al., 1989; Rosas and Bliss, 1986; Rosas et al., 1987). The bean germplasm selection methods developed in Latin America and U.S.A. have already shown the feasibility of breeding superior lines for N_2 fixation traits (Bliss, 1985; McFerson et al., 1982; Rosas et al., 1987). The high variability of fixation among bean cultivars and the need to apply N fertilizers to obtain high yields particularly in mixed cropping systems, are important reasons to study the potential of beans so that superior cultivars which maintain high rates of N_2 fixation in fertile, nitrogen-rich soils can be identified (Atkins, 1984).

The purpose of this study was to evaluate superior cultivars grown under three levels of N fertilizer in a maize-bean association for symbiotic traits. Estimates of N_2 fixation were made using the natural variation of ^{15}N to determine the amount of N in the plant coming from fixation (Hardarson, 1993).

Material and methods

Thirty-one bean lines previously selected by CNPAF-EMBRAPA (National Research Center for Rice and Bean-Goiania, GO, Brazil) for their ability to fix N_2, tolerate low phosphate levels, and produce high yields were grown in the field at the Experimental Station of the Escola Superior de Agricultura "Luiz de Queiroz". These included ten superior lines chosen from each of the following three improvement programs at CNPAF-EMBRAPA: a) a BNF programme, which has evaluated a total of 400 lines including the selected lines from the U.S.A. programme (Attewell and Bliss, 1985; Pereira et al., 1984a); b) a programme selecting for tolerance to low soil P, which has studied a total of 80 lines; c) and a programme selecting for high yielding bean cultivars for maize-bean associations, testing 40 climbing beans lines. The bean lines were: a) from BNF programme – CNF-480, Negro Argel, Venezuela-350, CNF-178, Carioca-80, BAT-76, Puebla-152, WBR21-58, WBR22-3, WBR22-34, WBR22-52, WBR22-50, WBR22-55 (Bliss et al., 1989), b) from the P-tolerance programme – 82-PVBZ-1771, 82-PVMX-1637, A-230, A-251, A-275, DOR-218, Rio Tibagi, NAG-24, NAG-52 and c) for maize-bean association – Preto Caruaru, Mulato Guanhães, Saula Santa Barbara, BSC-1, BSC-5, Cuva-168N, SPG-4, Parana-1 and Mexico-309. Carioca-80 was recommended for the last two groups.

The field experiment was planted in March 1986, at the beginning of the dry season in an Oxic Paleudalf soil previously planted with *Crotalaria juncea* at the Experimental Station of ESALQ-Piracicaba, state of S. Paulo, Brazil. The location of the site was 22°42′30″S latitude and 47°38′00″W longitude. During the experimental period, the mean average air temperature ranged from 21 to 25°C, with 152 mm per month mean rainfall. The plots were sprinkler-irrigated whenever the soil moisture tension reached 45 kPa. Simple superphosphate at the rate of 40 kg P_2O_{45} ha^{-1} was broadcast before planting. The chemical composition of the soil was: P (resin) 48 mg, O.M. (%) 3.5, pH (H_2O) 5.6, K, Ca and Mg 0.8, 3.0 and 1.5 meq 100 cm^{-3}, V(%) 55.

The hybrid maize variety Piranão was used as the bean companion plant in a 1:1 ratio combination, one row of maize and one row of bean. Population densities were 100,000 bean and 50,000 maize plants per hectare.

The experimental area was approximately 3240 m^2, with subplots of 9 m^2 (3 × 3 m). Each unit consisted of 4 rows of maize and 3 of bean per plot with 60 bean and 15 maize plants per 3 m row. Three rates of urea (0, 30 and 60 kg N ha^{-1}) were applied to the main plots and sub-plots with bean lines. Nitrogen was split in 15 kg N ha^{-1} and added at sowing and before flowering (28 days after emergence, DAE). Wheat cv. BH1146 (provided by Dr. A. Tulmann Neto, CENA-USP), which has a similar growth cycle as bean was used as the non-fixing reference plant. It was grown associated with maize in sub-plots in all blocks. The experiment was in a randomized split-plot design with 4 replications.

Bean seeds were pelleted with a peat-base inoculant of *Rhizobium leguminosarum* bv. *phaseoli* strain SEMIA-487 (IPAGRO, Porto Alegre, R5, Brazil), *Rhizobium tropici* strains CM-05 (CENA-USP, Piracicaba, SP, Brazil), and UMR-1899 (originally provided by Dr. P.

Graham, University of Minnesota, U.S.A.). Strain SEMIA-487 has not been characterized yet according to the new proposed classification for the root-nodule forming bacteria of common bean (Martinez-Romero et al., 1991; Segovia et al., 1993). Inoculants were prepared in CENA using gamma irradiated peat. Bean seeds were sprinkled with 10% (v:v) sucrose solution, pelleted with the mixed inocula containing approximately 10^9 *Rhizobium* cells per g peat and coated with calcium carbonate.

Ten bean plants from each sub-plot were collected for nodule mass, number and acetylene reduction activity (ARA) at 26 DAE and at 47 DAE. At physiological maturity (92 DAE), 10 plants from row 4 of each sub-plot were harvested, dried for 72 h at 60°C, weighed, ground to pass a 60 mesh sieve and analyzed for total N and ^{15}N abundance; the latter analysis being performed only on plants without N. The natural abundance determination of bean plants followed the Dumas combustion method and the $^{15}N/^{14}N$ ratios were determined in a Micromass 602E spectrometer, using the air natural ^{15}N abundance (0.3663 atom% ^{15}N) as reference.

Results

At 26 DAE, significant ($p < 0.05$) nitrogen (N) × line (L) interaction was found for nodule weight. At 47 DAE, significant differences were observed for both N levels and lines. Although nodulation traits were less expressive at this growth stage, plant weight and N varied in response to N addition and lines. At 92 DAE, N fertilizer affected N accumulation in the seed but not yield. However, line difference was the main factor influencing yield and accumulations of N and P in the seed. Analysis among the traits indicated that few variables were correlated. Exceptions were nodule weight vs. ARA at 26 DAE (r = 0.58 at $p < 0.05$); nodule weight at 26 DAE vs. N-seed derived from N_2 fixation (r = 0.36 at $p < 0.05$) and N-shoot at 47 DAE vs. N-seed both derived from N_2 fixation (0.55 at $p < 0.05$).

Nodulation traits varied significantly among lines at 26 DAE. At this stage, nitrogen (as low as 15 kg N ha^{-1}) was found to be detrimental to nodulation, affecting mostly the mass and activity of the nodules (Figs. 1a, 2). A wide range in nodule weight was observed primarily at the first sampling (Table 1). Mexico-309 was superior at 26 DAE, maintaining also high nodule weight at 47 DAE (data not shown). A-275, A-230, 82-PVBZ-1771, NAG-52, Parana-1 were the poorest lines for nodule mass formation in both samplings. Few differences were found among genotypes in ARA values, but these followed the same pattern as the nodulation parameters at 26 DAE (data not shown). The nodule activity of Mexico-309 at 47 DAE was low, perhaps indicating early senescence.

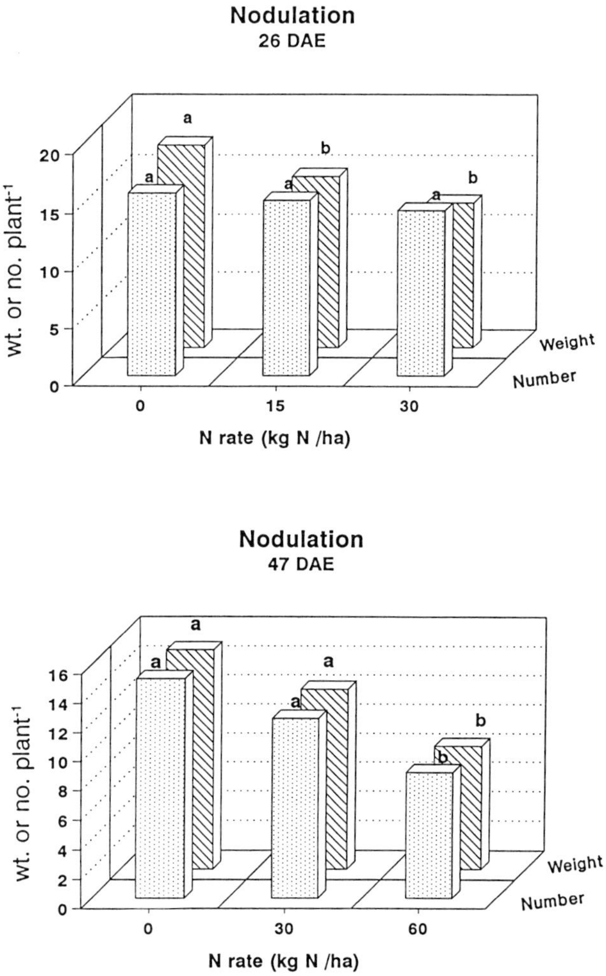

Fig. 1. Average nodule number and weight of 31 field-grown genotypes grown under three levels of N fertilizer in association with maize. **a** at 26 DAE and **b** at 47 DAE.

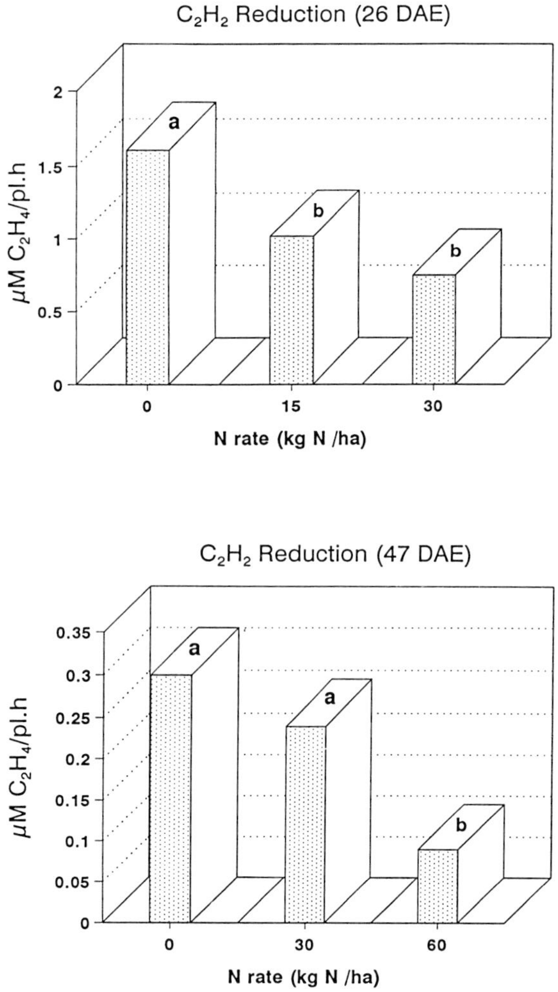

Fig. 2. Average acetylene reduction activity (ARA) of 31 field-grown bean genotypes grown under three levels of N fertilizer in association with maize, in two sampling times (26 and 47 DAE).

The decrease in nodule weight in response to increasing N levels was observed at 26 DAE and continued until flowering (Fig. 1a). At 47 DAE, N fertilizer also affected nodule number and nodule weight (Fig. 1b). At this stage, decreases were greatest at the 60 kg N ha^{-1} treatment, approximately 52% over the zero N treatment. At the early stage, only a few lines such as Mexico-309 and A-251 produced more nodule mass in presence of added N (Table 1). Among these, Mexico-309 produced the highest nodule weight in all three N treatments – 30.9 mg nodule per plant at 0 nitrogen, 34.2 mg at 30 kg N ha^{-1} and 21.4 mg at 60 kg N ha^{-1}.

Table 1. Nodule weight (mg/plant) in 31 field-grown bean genotypes determined at 26 DAE. Mean data of 4 replicates

Cultivar	N (kg ha^{-1})		
	0	30	60
82-PVMX-1637	20.3	24.2	7.7
BAT-76	22.4	10.0	10.9
BSC-1	15.3	9.4	7.4
82-PVBZ-1771	8.3	11.1	8.3
A-251	10.2	17.2	11.1
Preto Caruaru	11.9	9.4	12.2
CNF-480	28.4	17.1	17.3
Puebla-152	23.6	14.9	12.4
Mexico-309	30.9	34.2	21.4
Parana-1	8.8	7.4	11.5
WBR22-50	30.4	17.6	15.2
A-275	3.6	1.4	6.7
WBR22-58	15.3	11.0	12.2
CNF-178	18.2	9.3	11.8
Cuva-168N	11.9	11.0	14.9
NAG-52	14.9	12.6	5.3
DOR-218	13.2	7.8	19.3
SPG-4	12.4	12.6	12.2
M. Guanhaes	10.7	12.1	9.3
Carioca-80	18.2	13.8	13.9
NAG-24	20.1	16.1	12.7
Ven.-350	26.1	15.4	12.5
Negro Arg	25.0	25.2	15.5
Saula Sta. Barb.	23.7	15.1	14.0
WBR22-3	19.5	12.8	18.8
BSC-5	16.1	15.8	13.2
A-230	3.2	12.2	3.8
WBR22-34	16.2	19.9	12.5
WBR22-52	22.3	15.2	10.0
WBR22-55	19.7	14.0	11.2
Rio Tibagi	10.5	6.3	16.1

L.S.D. ($p < 0.05$) = 18.9

Nodule weights for WBR22-50, Negro Argel, Venezuela-350 and CNF-480 averaged 25–28 mg per plant, for 0 N ha^{-1} treatment. These lines and WBR22-3, WBR22-34, WBR22-52, WBR22-55 had the largest nodule weight. Lines 82-PVBZ-1771, A-230 and A-275 produced the least nodule weights.

The ARA varied significantly with N fertilizer level and plant age (Fig. 2). At 26 DAE, ARA rates were 37% and 52% less, when 15 and 30 kg N ha^{-1} were applied respectively. This effect was also observed at 47 DAE, when ARA rates were 20% and 73% lower for the 30 and 60 vs. 0 N treatments. Although nodule number did not decrease significantly and nodule weight decreased only 19% and 28% at the initial 15 and 30 kg N ha^{-1} additions (when only half of the N was added), nitrogenase activities of the nodules were sensitive to the presence of N fertilizer. Correlations between ARA and nodule weight in treatments without N were r = 0.918** ($p < 0.01$) at 26 DAE and $r = 0.556$* ($p < 0.05$) at 47 DAE. At 47 DAE, N_2 fixing activity was depressed in all treatments, with lowest activity observed under the highest N rate. The correlation between ARA and nodulation at this stage was lower than at 26 DAE. Most of the nodules were senescent, indicating that the bean plants were under stress perhaps from shading by the maize plants and/or a virus disease in some of the bean plants.

N fertilizer did not significantly increase bean yield, but enhanced accumulation of N in the shoot and seeds, particularly at the highest N-rate (Table 2). Almost all lines showed a high dependence on exogenous N, but some were superior at 0 kg N ha^{-1} (Table 3). Venezuela-350 had the highest total N accumulation, followed by NAG-24 and WBR22-34 (data not shown). Preto Caruaru, WBR22-55, WBR22-52, NAG-24, DOR-218, CNF-178 and WBR22-34 were ranked the highest with respect to seed yield at N_0. At 30 kg N ha^{-1}, CUVA-168N, NAG-24, WBR22-34 and WBR22-52 were superior and at 60 kg N ha^{-1}, WBR22-50 was the superior one. The WBR group (WBR22-3, WBR22-34, WBR22-50, WBR22-52, WBR22-55) gave good average yields in all N treatments.

NAG-24 and Venezuela-350 had the highest shoot weight at 47 DAE, averaging 5.57 and 5.40 g per plant (Table 3). Highest seed yields among lines were produced by WBR22-50, followed by NAG-24, WBR22-55, WBR22-34 and WBR22-52. The high nodule weight-producing lines – Puebla-152, Venezuela-350, Negro Argel, CNF-480 and Mexico-309 – were not necessary high seed producers, with averages of 1.2 to 2.3 g seeds per plant. WBR22-50 and NAG-24 were the most efficient in accumulating N, perhaps due to the higher seed yields. NAG-24 also had high values for shoot weight (5.57 g/plant), seed yield (2.73 g/plant) and N-seed accumulation (122.6 mg N/plant).

The distributions of lines according to proportion of nitrogen derived from N_2 fixation (%Ndfa) in the plant at 47 DAE and the proportion of %Ndfa in the seed N at 92 DAE are shown in Fig. 3. There was a positive correlation (r = 0.55* at $p < 0.05$) between those two traits.

Table 2. Average weight and N accumulated in the shoot at 47 DAE; seed weight and N at 92 DAE in 31 genotypes in response to N additions. Mean data of 124 replicates

N treatment (kg ha^{-1})	Shoot		Seed	
	Weight[a]	Total N[b]	Weight[a]	Total N[b]
0	4.17	131	1.98	88
30	4.36	134	1.95	88
60	4.68	156	2.17	100
L.S.D. ($p < 0.05$)	0.41	10.2	0.09	10.1

[a] g per plant.
[b] mg N per plant.

Table 3. Shoot weight (47 DAE), yield and seed-N accumulation (92 DAE) in 31 bean genotypes

Cultivars	Shoot (g/plant)	Seed	
		g/plant	(mg N/plant)
82-PVMX-1637	4.15	1.79	86.5
BAT-76	4.48	2.35	103.1
BSC-1	4.78	2.53	112.7
82-PVBZ-1771	3.06	0.50	23.6
A-251	5.06	2.36	102.2
Preto Caruaru	5.25	2.52	114.9
CNF-480	4.28	2.09	91.1
Puebla-152	3.93	1.18	50.5
Mexico-309	4.03	1.48	68.5
Parana-1	2.45	1.44	59.5
WBR22-50	4.51	2.96	134.2
A-275	2.33	1.56	52.2
WBR22-58	5.02	2.48	108.0
CNF-178	3.95	2.32	104.1
Cuva-168N	4.18	2.56	114.0
NAG-52	4.15	2.23	106.0
DOR-218	3.70	2.15	98.0
SPG-4	4.02	2.04	93.1
M. Guanhaes	3.47	1.91	79.8
Carioca-80	3.87	1.40	70.3
NAG-24	5.57	2.73	122.6
Ven.-350PS	5.40	2.11	98.9
Negro Argel	4.28	1.94	94.2
Saula Sta. Barb.	4.22	1.64	74.9
WBR22-3	4.23	2.39	107.6
BSC-5	5.16	1.71	78.7
A-230	3.76	1.36	63.3
WBR22-34	5.03	2.68	123.3
WBR22-52	4.69	2.56	116.6
WBR22-55	4.19	2.68	110.9
Rio Tibagi	3.75	1.75	83.2
L.S.D. ($p < 0.05$) =	1.85	1.37	54.3

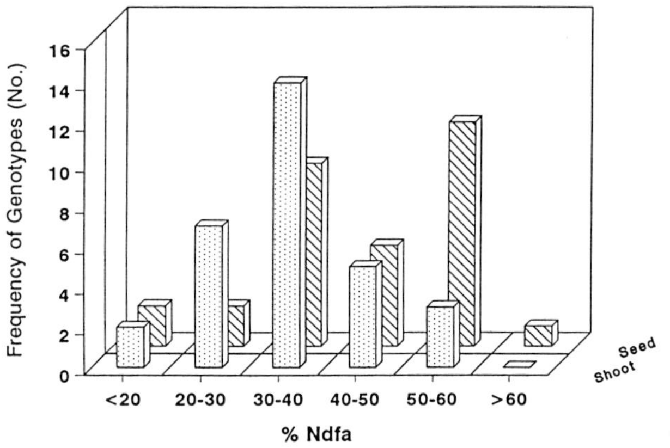

Fig. 3. Frequency of lines for their ability to fix N_2 (%Ndfa) in the bean shoot at 47 DAE and accumulate N_2 fixed in the seeds at 92 DAE.

CNF-178, Mexico-309, WBR22-3, CNF-480, Puebla-152 and Venezuela-350 consistently gave highest amounts of fixed N_2 (Table 4). An efficient conversion of this form of N from vegetative to reproductive parts occurred in CNF-480, Puebla-152, Venezuela-350, BSC-1, Saula Santa Barbara, A-251, WBR22-3, Negro Argel and WBR22-55, all of which had more than 50% of N in the seed from N_2 fixation.

When comparing the proportion of N_2 fixed in both shoots and seeds, several lines had values below the average of 30%. Also, the lines varied in their ability to convert N from shoot to seeds. CNF-480, CNF-178, BSC-1, Puebla-152, A-251, WBR22-55, Venezuela-350, Negro Argel, and WB22-3 converted more than 50% of the shoot N to seed N, with a maximum value of 60.6% for CNF-480. Carioca-80 showed low N_2 fixation with a proportion of 23.2% in the shoot and 19.1% in the seeds and a low conversion of N from shoot to seed, as was found previously by Ruschel et al. (1982). This pattern of N accumulation by N_2 fixing beans has also been demonstrated for other bean plants grown in temperate regions (Cornet et al., 1980, Westermann et al., 1981) and in soybean (Ronis et al., 1985).

Estimates of the fixed N_2 in the shoot (47 DAE) gave values from 5.6 to 21.1 kg N ha^{-1} and in the seeds (92 DAE) from 2.2 to 18.6 kg N ha^{-1} (Table 5). Although low, two main groups can be distinguished among lines. Several good N_2 fixers were listed among the first group for both shoot and seed data.

Table 4. Percentage of N in shoot (47 DAE) and seed (92 DAE) derived from atmosphere for 30* bean genotypes grown in association with maize

N derived from atmosphere (%)			
Line	Shoot	Line	Seed
CNF-178	56.6 (12.6)**	CNF-480	60.6 (9.7)
Mex-309	52.2 (8.9)	Puebla-152	59.6 (9.7)
WBR22-3	50.5 (10.9)	BSC-1	59.6 (9.7)
CNF-480	49.5 (6.5)	CNF-178	58.5 (9.7)
MULATO G.	48.5 (10.0)	A-251	55.3 (7.7)
PUEB-152	47.5 (8.2)	VEN.-350	54.3 (11.9)
VEN.-350	46.5 (13.8)	WBR22-3	54.3 (14.0)
WBR22-50	45.4 (8.3)	N. ARGEL	53.2 (18.2)
WBR22-52	39.4 (8.6)	PARANA-1	51.1 (6.7)
BSC-1	39.4 (10.3)	WBR22-55	51.1 (7.7)
SAULA SB	39.4 (8.6)	MEX-309	50.0 (14.0)
BAT-76	38.4 (15.0)	SAULA S.B.	50.0 (9.8)
A-251	36.4 (7.9)	WBR21-58	46.8 (14.0)
PARANA-1	36.4 (15.0)	WBR22-52	44.7 (9.9)
WBR21-58	36.4 (7.9)	CUVA-168N	43.6 (12.0)
WBR22-55	36.4 (9.6)	WBR22-34	43.6 (17.2)
N. ARGEL	35.4 (14.0)	BAT-76	42.6 (22.5)
WBR22-34	35.4 (13.2)	SPG-4	37.2 (17.2)
PR. CAR.	33.3 (10.6)	BSC-5	37.2 (8.9)
R. TIBAGI	31.3 (10.7)	WBR22-50	35.1 (4.2)
NAG-24	30.3 (15.2)	R. TIBAGI	35.1 (14.1)
82-PVBZ	28.3 (14.4)	82-PVBZ	34.0 (13.1)
CUV-168N	28.3 (17.2)	NAG-24	34.0 (14.1)
A-230	25.3 (7.9)	MULATO G.	33.0 (22.5)
82-PVMX	23.2 (11.0)	PRETO CAR.	30.9 (12.1)
CARIOC. 80	23.2 (10.2)	82-PVMX	30.9 (12.1)
SPG-4	22.4 (11.9)	DOR-218	23.4 (17.3)
NAG-52	21.1 (9.5)	A-230	20.2 (15.3)
DOR-518	18.2 (11.2)	CARIOCA-80	19.1 (9.2)
BSC-5	18.2 (11.2)	NAG-52	12.8 (9.3)

* Genotype A-275 was not included due to poor growth.
** Numbers in parenthesis represent standard deviation of each mean.

Discussion

Numerous studies in Brazil have demonstrated the variability in yield response to addition of N fertilizer in common bean (Almeida et al., 1973; Eira et al., 1973; Fontes et al., 1965; Miyasaka et al., 1964, 1966). Although N can increase yields, the amounts required for maximum yield are costly. Usually beans are cultivated as a subsistence crop and often have widely fluctuating prices in the market.

Data from other studies indicate that the major constraints to increased N_2 fixation and high bean yields in Latin America are primarily poor agricultural management – lack of irrigation, phosphorous and lime (Bonetti et al., 1984;

Table 5. Amounts (kg N ha^{-1}) of N derived from atmosphere estimated in the shoot at 47 DAE and in the seed (92 DAE) of 30* bean genotypes intercropped with maize plants

N derived from atmosphere (kg N ha^{-1})			
Cultivar	Shoot	Cultivar	Seed
CNF-480	21.1	CNF-178	18.6
VENEZUELA-350	19.8	A-251	16.5
CNF-178	17.6	BSC-1	16.1
PRETO CARUARU	17.5	WBR22-55	16.1
MEXICO-309	17.4	WBR22-34	14.8
WBR22-50	17.4	WBR22-52	14.7
MULATO GUANHAES	16.7	CUVA-178N	13.6
NAG-24	16.3	PRETO CARUARU	13.4
WBR22-3	15.7	VENEZUELA-350	12.9
BAT-76	15.4	CNF-480	12.6
WBR21-58	15.1	NEGRO ARGEL	12.5
A-251	14.8	WBR22-3	12.2
PUEBLA-152	14.6	PARANA-1	11.6
BSC-1	14.6	BAT-76	10.8
SAULA STA. B.	13.1	WBR22-50	10.8
WBR22 34	13.1	NAG-24	10.8
WBR22-52	12.4	WBR21-58	9.8
WBR22-55	11.5	PUEBLA-152	9.0
NEGRO ARGEL	11.5	SAULA STA. B	9.0
CARIOCA-80	11.3	SPG-4	9.0
CUVA-168N	11.0	MEXICO-309	8.8
RIO TIBAGI	10.1	BSC-5	7.6
PARANA-1	9.0	82-PVMX 1637	7.2
A-230	9.0	RIO TIBAGI	7.0
SPG-4	8.4	MULATO GUANHAES	6.8
NAG-52	8.2	DOR-218	4.4
82-PVBZ 1771	7.9	NAG-52	3.7
BSC-5	7.8	CARIOCA-80	3.4
82-PVMX 1637	7.3	A-230	3.0
DOR-218	5.6	82-PVBZ 1771	2.2

* Genotype A-275 was not included due to poor growth.

Saito and Ruschel, 1978; Saito et al., 1983). Saito and Ruschel (1978) have indicated that, once these factors are corrected, the naturally established microsymbiont *Rhizobium* from traditional bean fields usually is adequate for reasonable amounts of N_2.

In the present study, the symbiotic response to *Rhizobium* by the 31 lines showed that the higher N levels in the shoot due to N fertilizer were not sufficient to increase yield. N fertilization affected nodule formation and functioning (ARA), however our results reinforce the need to identify appropriate rates of N application that will not suppress nodulation and N_2 fixation. In the particular case of the maize-bean intercropping, large amounts of N fertilizer applied to the companion maize plant would also suppress newly emerging nodules on the lateral roots which have been found to be especially important sites for BNF in grain legumes (Hardarson et al., 1989; Wadisirisuk et al., 1989; Wolyn et al., 1989). The performance of several lines in the maize-bean association was superior compared to a monocrop system, especially when bean is introduced after harvesting the corn (Pereira et al., 1984b). In a monocrop situation, Carioca was superior to Negro Argel, Venezuela-350 and Rio Tibagi (Duque et al., 1985).

The lines in this experiment showed negative correlation between symbiotic activity and seed yield when intercropped with maize. Other authors have found similar results in a bean monocrop (Franco and Döbereiner, 1967; Pessanha et al., 1972). Several papers report the preferential translocation of fixed N_2 from nodules directly to flowers/fruits of soybean (Herridge, 1982; Neves et al., 1985; Pate et al., 1980) and beans (Thomas and Hungria, 1988). Although a difficult task, it may be possible to select bean cultivars for both high N_2 fixation and high seed yield through plant breeding (St. Clair et al., 1988; St. Clair and Bliss, 1991; Westermann and Kolar, 1978). The preferential accumulation of fixed N_2 in the seeds determined in most of the high N_2 fixing materials and also found by Miranda and Bliss (1991), appears promising from future recommendations on breeding beans for N_2 fixation.

None of the lines examined were outstanding for all traits. Mexico-309 was superior for nodulation parameters but poor for seed yield, while Preto Caruaru produced high seed yield, but was inferior in nodulation traits. However, Negro Argel and CNF-178 were quite good for all seven studied parameters – ARA, nodule number and mass, total N in shoot, seed yield, N_2 fixed in the shoot and seed. Carioca-80, a widely planted cultivar, was a poor fixer under this intercropping system with maize.

Recent progress on bean improvement conducted at CNPAF-EMBRAPA (Brazil), CIAT (Colombia) and in U.S.A. has already demonstrated the feasibility of increasing N_2 fixation and yield in common bean by breeding and selecting progenies with high ARA and seed yields (McFerson et al., 1982; Rosas and Bliss, 1986; Rosas et al., 1987; St. Clair et al., 1987; St. Clair and Bliss, 1991; Pereira, 1992). Selection for grain yield has shown that larger genotypic differences can be detected under the competitive environment of intercropping (Zimmermann et al., 1984) than in sole bean crop. Under this condition, selection for N_2 fixation may play an important role for increasing bean yield.

Acknowledgements

The authors wish to acknowledge the support for this research which was received from the U.S. National Academy of Science (BNF-BR-4-86-53) and Joint FAO/IAEA Division (311-D1-BRA-4249). They would also like to thank R S Pacovsky, M Fried, T Muraoka, K Reichardt, P Libardi, for their assistance and scientific support; and M Bizetto and F Montrazzi for technical assistance. S.M.T. acknowledges fundings from Conselho Nacional de Desenvolvimento Cientifico e Tecnologico (CNPq).

References

Almeida D L, Pessanha G G and Penteado A F 1973 Effects of liming and phosphorus and nitrogen fertilizer on nodulation and yield of beans (*Phaseolus vulgaris*). Pesqui. Agropecu. Bras. 8, 127–130.

Attewell J and Bliss F A 1985 Host plant characteristics of common bean lines selected using indirect measures of N_2 fixation. *In* Nitrogen Fixation Research Progress. Eds. H J Evans, P J Bottomley and W E Newton. pp 3–9. Martinus Nijhoff Publishers, Dordrecht.

Atkins C A 1984 Efficiencies and inefficiencies in the legume/*Rhizobium* symbiosis: A review. Plant and Soil 82, 273–284.

Bliss F A 1985 Breeding for enhanced dinitrogen fixation potential of common bean (*Phaseolus vulgaris* L.). *In* Nitrogen fixation and CO_2 metabolism. Eds. P W Ludden and J E Burris. pp 303–310. Proc. 14th Steenboock Symposium, University of Wisconsin, Madison, WI. 17–22 June 1984. Elsevier, New York.

Bliss F A, Pereira P A A, Araujo R S, Henson R A, Kmiecik K A, McFerson J R, Teixeira M G and Da Silva C C 1989 Registration of five high nitrogen fixing common bean germplasm lines. Crop Sci. 29, 240–241.

Bonetti R, Montanheiro M N S and Saito S M T 1984 The effects of phosphate and soil moisture on the nodulation and growth of *Phaseolus vulgaris*. J. Agric. Sc. 103, 95–102.

Cornet D, Meulemans M and Otoul E 1980. Une association fixatrice d'azote prometteuse: la symbiose haricot de Lima (*Phaseolus lunatus* L.) – *Rhizobium*. Bull. Rech. Agron. Gembloux 15, 17–30.

DuBois J D and Burris R H 1986 Comparative study of N uptake and distribution in three lines of common bean (*Phaseolus vulgaris* L.) at early pod filling stage. Plant and Soil 93, 79–86.

Duque F F, Neves M C P, Franco A A, Victoria R L and Boddey R M 1985 The response of field-grown *Phaseolus vulgaris* L. to *Rhizobium* inoculation and the quantification of N_2 fixation using ^{15}N. Plant and Soil 88, 333–343.

Eira P A, Pessanha G G, Britto D P P S and Carbajal A R 1973 Comparison of experimental design in a field trial on fertilization of black beans (*Phaseolus vulgaris*) with mineral nitrogen and phosphorus. Pesqui. Agropecu. Bras. 8, 121–125.

Fontes L A N, Gomes F R and Vieira C 1965 Effect of lime, nitrogen, phosphorus and sulphur on dry beans of the Zona da Mata, Minas Gerais State. Ceres 12, 265–285.

Franco A A and Döbereiner J 1967 Especificidade hospedeira na simbiose com Rhizobium-feijao e influencia de diferentes nutrientes. Pesqui. Agropecu. Bras. 2, 467–474.

Graham P H 1981 Some problems of nodulation and symbiotic nitrogen fixation in *Phaseolus vulgaris* L.: A review. Field Crops Res. 4, 93–112.

Graham P H and Halliday J 1977 Inoculation and nitrogen fixation in the genus *Phaseolus*. *In* Exploiting the Legume-*Rhizobium* Symbiosis in Tropical Agriculture, pp 313–334. Univ. Hawaii Coll. Tropic. Agric. Misc. Publ. 145.

Graham P H and Temple S R 1984 Selection for improved nitrogen fixation in *Glycine max* (L.) Merr. and *Phaseolus vulgaris* L. Plant and Soil 82, 315–327.

Hardarson G and Danso S K A 1993 Methods for measuring biological nitrogen fixation in grain legumes. Plant and Soil 152, 19–23

Hardarson G, Golbs M and Danso S K A 1989 Nitrogen fixation in soybean (*Glycine max* L. Merrill) as affected by nodulation patterns. Soil Biol. Biochem. 21, 783–787.

Herridge D F 1982 Use of the ureide technique to describe the nitrogen economy of field grown soybeans. Plant Physiol. 70, 7–11.

Martinez-Romero E, Segovia L, Mercante F M, Franco A A, Graham P H and Pardo M A 1991 *Rhizobium tropici*, a novel species nodulating *Phaseolus vulgaris* L. bean and Leucaena sp trees. Int. J. Syst. Bacteriol. 41, 417–426.

McFerson J F, Bliss F A and Rosas J C 1982 Selection for enhanced nitrogen fixation in common beans, *Phaseolus vulgaris* L. *In* BNF Technology for Tropical Agriculture. Eds. P H Graham and S C Harris. pp 39–44. CIAT, Cali, Colombia.

Miranda B D and Bliss F A 1991 Selection for increased seed nitrogen accumulation in common bean: Implications for improving dinitrogen fixation and seed yield. Pl. Breeding Z. Pflanzenfz. 106, 301–311.

Miyasaka S, Freire E S and Mascarenhas H A A 1964 Fertilizer experiments with soybeans and dry beans in 'Arenito Botucatu' soil. Bragantia 23, 45–54.

Miyasaka S, Freire E S, Igue I, Schmidt N C and Leite N 1966 Mineral fertilizers for dry beans. V. Effects of nitrogen, phosphorus, potassium, sulphur and micronutrients mixture in two soil of the Paraiba Valley. Bragantia 25, 307–316.

Neves M C P, Didonet A D, Duque F F and Döbereiner J 1985 *Rhizobium* strain effects on nitrogen transport and distribution in soybeans. J. Exp. Bot. 36, 1179–1192.

Pate D F 1982 Use of the ureide technique to describe the nitrogen economy of field grown soybeans. Plant Physiol. 70, 7–11.

Pereira P A A, Araujo R S, Moreira R G and Steinmetz S 1984a Capacidade de genotipos de feijoeiro de fixar N_2 atmosferico, Pesqui. Agropecu. Bras. 19, 811–815.

Pereira P A A, Rocha R E M and Araujo R S 1984b Nodulacao do feijoeiro em monocultivo e associado com fileiras de milho maduro. Pesqui. Agropecu. Bras. 19, 951–954.

Pereira P A A, Burris R H and Bliss F A 1989 ^{15}N-dinitrogen fixation potential of genetically diverse bean lines (*Phaseolus vulgaris* L.). Plant and Soil 120, 171–179.

Pessanha G G, Franco A A, Döbereiner J, Groszman A and Britto D P S 1972 Negative correlation between nodulation and seed production in bean (*Phaseolus vulgaris* L.) on soils in which nitrogen is not a limiting factor. Pesqui. Agropecu. Bras. 7, 49–56.

Rennie R J and Kemp G A 1983a Nitrogen fixation in field beans (*Phaseolus vulgaris*) quantified by nitrogen-15 isotope dilution. I. Effect of strains of *Rhizobium phaseoli*. Agron. J. 75, 640–644.

Rennie R J and Kemp G A 1983b Nitrogen fixation in field beans (*Phaseolus vulgaris*) quantified by nitrogen-15 isotope dilution. II. Effect of cultivars of beans. Agron. J. 75, 645–649.

Rennie R J and Kemp G A 1984 ^{15}N-determined time course for N_2 fixation in two cultivars of field bean. Agron. J. 76, 146–154.

Ronis D H, Sammons D J, Kenworthy W J and Meisinger J J 1985 Heritability of total and fixed N content of the seed in two soybean populations. Crop Sci. 25, 1–4.

Rosas J C and Bliss F A 1986 Improvement of the nitrogen fixation potential of common beans in Latin America. CEIBA 27, 245–259.

Rosas J C, Kipe-Nolt J, Henson R A and Bliss F A 1987 Estrategias de mejoramiento para incrementar la capacidad de fijacion biologica de nitogeno del frijol comun en America Latina. CEIBA 28, 39–57.

Ruschel A P, Vose P B, Matsui E, Victoria R L and Saito S M T 1982 Field evaluation of nitrogen fixation and nitrogen utilization by *Phaseolus vulgaris* bean varieties determined by nitrogen-15 isotope dilution. Plant and Soil 65, 397–408.

Saito S M T 1982 Field evaluation of the symbiotic fixation abilities among *Rhizobium phaseoli* strains. Pesqui. Agropecu. Bras. 17, 999–1006.

Saito S M T and Ruschel A P 1978 Influence of lime, phosphorus fertilizer and micronutrients in the natural nodulation of beans (*Phaseolus vulgaris*). An. Esc. Sup. Agric. "Luiz de Queiroz" 25, 545–556.

Saito S M T, Martins E C S, Freitas J R and Roston A J 1983 Natural occurrence of mycorrhiza and *Rhizobium phaseoli* in areas under bean cultivation. Pesqui Agropec. Bras. 18, 855–861.

St. Clair D A and Blis F A 1991 Intrapopulation recombination for N-15-determined dinitrogen fixation ability in common bean. Pl. Breeding. Z. Pflanzenfz. 106, 215–225.

St. Clair D A, Wolyn D J, DuBois J, Burris R H and Bliss F A 1988 Field comparison of dinitrogen fixation determined with nitrogen-15-depleted and nitrogen-15-enriched ammonium sulfate in selected inbred backcross lines of common bean. Crop Sci. 28, 773–778.

Segovia L, Young J P W and Martínez-Romero E 1993 Reclassification of American *Rhizobium leguminosarum* bv. *phaseoli* type I strains as *Rhizobium etli* sp. nov. Int. J. Syst. Bacteriol. (In press).

Wadisirisuk P, Danso S K A, Hardarson G and Bowen G D 1989 Influence of *Bradyrhizobium japonicum* location and movement on nodulation and nitrogen fixation in soybeans. Appl. Envir. Microbiol 55, 1711–1716.

Westermann D T and Kolar J J 1978 Symbiotic $N_2(C_2H_2)$ fixation by bean. Crop Sci. 18, 986–990.

Westermann D T, Kleinkopf G E, Porter L K and Leggett G E 1981 Nitrogen sources for bean seed production. Agron. J. 73, 660–664.

Westermann D T, Porter L K and O'Deen W A 1985 Nitrogen partitioning and mobilization patterns in bean plants. Crop Sci. 25, 225–228.

Wolyn D J, Attewell J, Ludden P W and Bliss F A 1989 Indirect measures of N_2 fixation in common bean (*Phaseolus vulgaris* L.) under field conditions: The role of lateral root nodules. Plant and Soil 113, 181–187.

Zimmermann M J, Rosielle A A and Waines J G 1984 Heritabilities of grain yield of common bean in sole crop and in intercrop with maize. Crop Sci. 24, 641–644.

Plant and Soil **152**: 103–106, 1993.

PLSO BNF10

Nitrogen fixation in breeding lines of *Phaseolus vulgaris* L.

J.A. KIPE-NOLT[1], H. VARGAS[1] and K.E. GILLER[2]
[1]*Centro Internacional de Agricultura Tropical (CIAT), Apartado Aéreo 6713, Cali, Colombia;*
[2]*Department of Biochemistry, Physiology and Soil Science, Wye College, University of London, Wye, Nr. Ashford, Kent, TN25 5AH, UK*

Key words: acetylene reduction, ^{15}N isotope, nitrogen fixation, nodulation, *Phaseolus vulgaris* L.

Abstract

Nitrogen fixation in nine common bean (*Phaseolus vulgaris* L.) lines was estimated using the ^{15}N isotope dilution method at two locations in two seasons. In the first season at one location no N_2 fixation was detected while in the second season up to 51 kg N ha^{-1} were estimated. There were significant differences between lines and correlations between trials were significant for the amounts of N_2 fixed, but not for total shoot nitrogen. The plants that fixed the most nitrogen nodulated rapidly after germination. Differences in maximum nodule mass, but not specific nodule activity, were detected also.

Introduction

Nitrogen deficiency limits productivity in many bean growing areas as indicated by the positive responses to nitrogenous fertilizers. The recommendations in both tropical and temperate areas are that farmers apply at least minimal doses of N. Research directed toward increasing N_2 fixation to overcome this constraint in a cheaper and more sustainable way, has emphasized improving both the plant genotype and the rhizobial strain components of the symbiosis.

A plant breeding programme to enhance N_2 fixation in small-seeded bush beans adapted primarily for Central America was initiated at the International Center for Tropical Agriculture (CIAT) in the late seventies (CIAT, 1978). Lines, coded RIZ, were selected based on performance for a range of nitrogen fixation-associated characters in three field sites and in the glasshouse (CIAT, 1987). Bean genotypes frequently demonstrate a fairly narrow range of environmental adaptation, in addition to temperature-photoperiod interactions. Therefore a better understanding of variation in genotype ranking for N_2 fixation across locations and years would assist in designing an efficient evaluation and selection scheme.

In this study N_2 fixation in nine bean lines was compared at two locations, having contrasting soil but similar climatic conditions, and during two seasons. The ^{15}N isotope dilution method was used to estimate nitrogen fixation. Nodulation, acetylene reduction activity and nitrogen accumulation were monitored throughout the growing cycle.

Materials and methods

Five bean lines that had been selected for N_2 fixation-associated characters (RIZ lines) and four lines from other breeding programmes were evaluated in four experiments. The dwarf sorghum cultivar SC326-6, was used as a non-fixing control plant in all four experiments. A non-nodulating soybean variety was also included in one experiment. All the bean lines have similar growth habit (indeterminate bush), seed size and time to maturity. Granular inoculum containing three highly effective *Rhizobium leguminosarum* bv. *phaseoli* strains was added in the furrow at sowing at a rate of 1 g m^{-1}. Phosphorus, potassium and micronutrient fertilizers were added for optimum growth. Insecticides and fungicides were applied as needed throughout the season.

Experiments were conducted at the CIAT research centre, Palmira, Colombia in the first and second semesters 1985. The centre is at 1000 m.a.s.l. and the soil type is isohyperthermic Aquic Hapludoll. The other two experiments, also conducted during the 1985 seasons, were at the Santander de Quilichao station, 42 km from Palmira at the same altitude in an isohyperthermic Plinthidic Kandiudox soil. Ten plant genotype treatments in 2 m × 6.5 m plots were randomized within six replicate blocks. Isotope labelled fertilizer, 10 kg N ha^{-1} of $(NH_4)_2SO_4$ (15 atom % ^{15}N excess), was applied to miniplots (2 m × 1 m) within each of the main treatment plots. Unlabelled fertilizer was applied to the remainder of the plot. The nitrogen fertilizer was added to a sucrose solution calculated to give a C:N ratio of 10:1, and watered on to each plot one week prior to sowing. Rows were 50 cm apart (4 per plot) and seeds were spaced at 10 cm within the rows.

The central 0.8 m^2 of the isotope-labelled plots were harvested when the bean plants reached the late pod-filling stage, before leaf fall (63 days after sowing in Palmira, and 70 and 64 days after sowing in the two Santander seasons, respectively). Samples were dried, weighed and ground. Nitrogen content was determined by Kjeldahl digestion and an automated indophenol-blue method. Nitrogen in the digests was concentrated by a Conway microdiffusion technique (Conway, 1939) and ^{15}N enrichments were measured using a Micromass 622 mass spectrometer. Fertilizer recovery and nitrogen fixation were calculated using the equations of Witty (1983).

Samples of 10 plants were taken from the unlabelled portion of the plots at 8 to 12 day intervals between 14 and 60 days after sowing. Root systems were excavated carefully, separated from the shoots and immediately put in sealable bottles into which 10% acetylene was injected. Gas samples were taken after 10 min and 1 hr and ethylene concentration determined using a Perkin-Elmer gas chromatograph with flame ionization detector and Poropak N column. Nodules were separated, counted, dried and weighed. Roots, shoots and pods were dried, weighed and analyzed separately for N content. A final harvest (6 m^2) was made at between 84 and 90 days after sowing and grain yield estimated.

Results and discussion

In the first semester in Palmira no fixation was detected by any of the bean genotypes using the ^{15}N isotope dilution method. Nodulation was poor (mean maximum nodule number 10/plant), as was acetylene reduction activity. High nitrogen mineralization rates may have been partly responsible, however very low grain yields in that experiment indicate that other factors also contributed to the very low fixation.

Table 1. Total shoot N and N_2 fixation in nine bean lines evaluated in three experiments: S85A (Santander 1985, first semester); S85B (Santander 1985, second semester) and P85B (Palmira 1985, second semester)

Line	N_2 fixed (kg ha^{-1})			%N_2 from fixation			Total N (kg ha^{-1})		
	S85A	S85B	P85B	S85A	S85B	P85B	S85A	S85B	P85B
RIZ32	23	35	50	37	33	43	62	106	116
RIZ68	23	37	36	38	40	50	61	93	72
BAT477	21	41	51	35	40	49	60	103	104
BAT1554	21	27	36	30	37	39	70	73	92
DOR41	21	29	27	34	32	37	62	91	73
RIZ44	20	24	23	37	34	33	54	71	70
RIZ53	19	9*	39	33	24*	37	58	38*	105
RIZ30	17	29	31	31	39	40	55	74	78
WTE3	14	18	18	26	30	19	54	60	95
SE	2.7	6.7	7.3	2.7	3.6	8.3	4.8	11	15

* Severely damaged by virus.
SE: Standard error of mean difference.

Nitrogen fixation in the nine bean lines in the other three experiments is shown in Table 1. In the first semester in Santander the amount of nitrogen fixed ranged from 14 to 23 kg N ha^{-1} and differences between the highest and lowest fixing lines were significant. In the second semester in Santander fixation levels were generally higher with the best line, BAT477, fixing 41 kg N ha^{-1}. The same three lines that had shown largest amounts of fixation in the first semester were good in the second season; BAT477, RIZ32 and RIZ68. In contrast to the first semester in Palmira, fixation levels were high in the second semester, ranging from 18 to 51 kg N ha^{-1}. BAT477, RIZ32 and RIZ68 were again high. No more than one-half of the total plant nitrogen of the nine lines in these locations was derived from fixation. RIZ68 showed highest percent nitrogen from fixation in all three trials, with values between 38 and 50%. WTE3 showed consistently low levels of fixation.

At Santander, second semester, a non-nodulating soybean was included as a non-fixing control in addition to the dwarf sorghum. Estimated levels of fixation were almost 40% lower using the soybean control ranging from 2 to 25 kg N ha^{-1}. The soybean established more quickly than the sorghum, taking up a greater proportion of its nitrogen early in the growth cycle, therefore the differences between the two control plants were more likely due to rooting pattern differences than to contrasting uptake patterns in a soil of declining ^{15}N enrichment. It was noted that the soybean tap root went considerably deeper into the soil than either the bean or sorghum roots (data not shown).

Grain yields of the nine lines in the three trials are shown in Table 2. Overall yields were good but slightly lower in the first semester at Santander. BAT477 and RIZ32 yielded well in all three trials. A significant ($p < 0.01$) linear correlation between amount of nitrogen fixed and yield was observed in that experiment ($r = 0.765$). Total shoot nitrogen was significantly correlated with yield in the Palmira experiment ($r = 0.968$).

Correlations between sites were significant for the amount of nitrogen fixed and for the % nitrogen from fixation except in the case comparing the two Santander trials (Table 3). However, there was no relationship between trials in total shoot nitrogen and for yield the correlation was significant only in the comparison of the two second semester trials. General adaptation to the conditions in both Palmira and Santander was a necessary characteristic for lines to be included in these trials so a correlation in yield was not surprising. Rainfall patterns varied between the two semesters in Santander and differences between genotypes in tolerance to the drought during flowering in the first semester may have contributed to the variable yield and total nitrogen responses. The more consistent rankings among lines for amount of N_2 fixed was interesting and suggests that preliminary selection at a single location could be effective. In previous studies with soybeans significant genotype by location interactions were not observed (Yoneyama et al., 1986; Snitwongse et al., 1989).

Table 2. Grain yield of nine bean lines evaluated in three experiments: S85A (Santander 1985, first semester); S85B (Santander 1985, second semester) and P85B (Palmira 1985, second semester)

Line	Yield (kg ha^{-1})		
	S85A	S85B	P85B
RIZ32	2350	3160	2950
RIZ68	2140	2260	2240
BAT477	2460	2760	2650
BAT1554	2110	3000	2700
DOR41	2340	1980	2210
RIZ44	2080	2830	2190
RIZ53	2170	2560	2890
RIZ30	2130	2570	2310
WTE3	1630	2690	2650
SE	130	180	190

SE: Standard error of mean difference.

The RIZ lines did not fix larger amounts of nitrogen than some lines selected in other breeding programmes. BAT477 is of particular interest since it is outstanding for drought tolerance. A more extensive evaluation of advanced/improved lines from other programmes is warranted and initially should take some priority over the evaluation of other germplasm.

Nodulation and acetylene reduction activity were monitored throughout the plant cycle. As in previous studies (Kipe-Nolt and Giller, 1993) no individual parameter was closely related to amounts of nitrogen fixed. Rankings among lines for three components of fixation, early nodula-

Table 3. Linear correlation coefficients of nitrogen fixation, total nitrogen and grain yield between the three experiments: S85A and S85B[a] (Santander 1985, first and second semesters respectively) and P85B (Palmira 1985, second semester)

	S85A vs. S85B	S85A vs. P85B	P85B vs. S85B
N fixed	0.739*	0.641*	0.879**
%N from fixation	0.339	0.710*	0.800**
Total N	0.362	0.200	0.401
Yield	−0.049	0.052	0.752*

*, ** Significant at $p < 0.05$ and $p < 0.01$ respectively.
[a] Not including RIZ53 that was affected by virus.

Table 4. Rankings among lines for early nodulation (14 days after sowing), maximum nodule mass (between 38 and 53 days after sowing) and specific nodule activity (nmoles C_2H_4 nodule^{-1} hr^{-1}, 38 days after sowing) of nine genotypes evaluated in 3 experiments[a]

Line	Early nodulation			Maximum nodulation			Specific nodule activity		
	S85A	S85B	P85B	S85A	S85B	P85B	S85A	S85B	P85B
RIZ32	1	1	8	5	3	7	8	5	3
RIZ68	6	2	2	8	2	3	1	1	6
BAT477	2	3	1	2	1	6	5	6	1
BAT1554	8	8	4	6	6	9	7	8	9
DOR41	4	7	5	4	7	4	6	3	5
RIZ44	9	6	7	3	9	1	9	9	8
RIZ53	3	4	9	7	4	5	3	7	2
RIZ30	5	5	3	1	5	2	4	2	4
WTE3	7	9	6	9	8	8	2	4	7

[a] Differences between lines in each experiment were significant ($p < 0.05$) for early nodulation and maximum nodule mass but not for specific nodule activity.

tion, maximum nodule mass and specific nodule activity are shown in Table 4. The three genotypes with highest levels of fixation RIZ32, RIZ68 and BAT477, all formed effective nodules rapidly after germination. RIZ30 and BAT477 were most consistent for high maximum nodule mass. Differences between genotypes in specific nodule activity (nmoles ethylene produced nodule^{-1}, hour^{-1}) were not significant in any of these trials. RIZ68, RIZ30 and RIZ53 tended to have high levels, but whether the lack of significant differences indicates that specific activity is not an important variable or whether it is just a problem of methodology (CVs for acetylene reduction activity were extremely high), requires further study.

Acknowledgements

We thank the staff of the bean microbiology section at CIAT, especially Aurora Duque, for their contribution to this work. The Rothamstead collaboration was supported by the Overseas Development Adminstration Research Scheme, R3648A.

References

CIAT 1978 Annual Report of the Bean Production Program, Cali, Colombia. 75 p.
CIAT 1987 Annual Report of the Bean Production Program, Cali, Colombia. 352 p.
Conway E J 1939 Microdiffusion analysis and volumetric error. Crosby-Lockwood, London.
Kipe-Nolt J A and Giller K E 1993 Plant and Soil 152, 107–114.
Snitwongse P, Siripaiboon C, Chaiwanakput P, Boonkerd N and Kucey R N 1989 Use of ARA and ^{15}N dilution techniques to measure N_2 fixation by soybean. Thai J. Agric. Sci. 21, 89-100.
Witty J F 1983 Estimating N_2-fixation in the field using ^{15}N labelled fertilizer: Some problems and solutions. Soil Biol. Biochem. 15, 631–699.
Yoneyama T, Nakano H, Kuwahara M, Takahashi T, Kambayashi I and Ishizuka J 1986 Natural ^{15}N abundance of field grown soybean grains harvested in various locations in Japan and estimate of the fractional contribution of nitrogen fixation. Soil Sci. Plant Nutr. 32, 443–449.

Plant and Soil **152**: 107–114, 1993.

PLSO BNF-11

A field evaluation using the ^{15}N isotope dilution method of lines of *Phaseolus vulgaris* L. bred for increased nitrogen fixation

J.A. KIPE-NOLT[1] and K.E. GILLER[2]
[1]*Centro Internacional de Agricultura Tropical (CIAT), Apartado Aéreo 6713, Cali, Colombia;*
[2]*Department of Biochemistry, Physiology and Soil Science, Wye College, University of London, Wye, Nr. Ashford, Kent, TN25 5AH, UK*

Key words: breeding, nitrogen fixation, nodulation, ^{15}N isotope, *Phaseolus vulgaris* L.

Abstract

N_2 fixation in lines of *Phaseolus vulgaris* was measured by ^{15}N-isotope dilution to determine whether a programme of crossing and recurrent selection had resulted in enhanced nitrogen fixation. In field experiments on an isohyperthermic Aquic Hapludoll soil the amount of N_2 fixed by the different lines ranged from 18 to 36 kg ha^{-1} (32 to 47% of plant N) in 56 days. The quantity of N_2 fixed and the proportion of plant N derived from fixation was not significantly greater in the lines selected for N_2 fixation (RIZ lines) than parental lines. Total shoot N ranged from 53 to 77 kg ha^{-1} and partitioning of N to pods differed from 28 to 52% among the lines which all had similar growth habit and duration. Nodulation patterns were also distinct. Nodules formed early (10 to 15 plant^{-1} at 13 days) in many lines, and smallest amounts of fixation were observed in those lines which nodulated slowly and did not form substantial nodule mass until after 40 days. The screening criteria used in the selection of the RIZ lines had been largely indirect with other factors such as disease resistance also being included. Progress for increasing N_2 fixation over good-fixing parental lines such as BAT76 was not significant and it is recommended that more attention be paid to early nodulation, to the use of soils with lower available N and to inter-crossing of lines having different good N_2 fixation traits in order to further enhance the potential for N_2 fixation in beans.

Introduction

Genetic variability for N_2 fixation-related characters has been demonstrated for both components of the legume – *Rhizobium* symbiosis. Many programmes have concentrated on rhizobial strain testing and selection as a means of improving N_2 fixation in the crop. Only in the last decade has much emphasis been placed on breeding and selection of the host legume to increase amounts of N_2 fixation. Legumes able to fix more nitrogen are especially important in developing countries where nitrogen nutrition is a serious constraint to greater crop productivity. With varying specific objectives and strategies, breeding and selection programmes have been established for improving N_2 fixation in alfalfa (Barnes et al., 1984), soybeans (Greder, 1983; Kueneman et al., 1984), clovers (Nutman, 1984), cowpea (Miller et al., 1986) and beans (CIAT, 1978, Graham and Temple, 1984; McFerson et al., 1982; McFerson, 1983).

The recurrent selection programme at CIAT originally centered on enhancing N_2 fixation in agronomically acceptable, indeterminate growth habit, small-seeded bush beans (*Phaseolus vulgaris* L.) (CIAT, 1978, 1984; Graham 1981). Throughout Central America and Brazil where these beans are grown, nitrogen availability frequently limits production. Criteria for selection of lines with greater potential for N_2 fixation included plant vigor, mass and number of nodules at mid-flowering, acetylene reduction activity, total-N content and grain yield of materials grown in both sand culture and in field trials with mild N-deficiency (Graham and Temple 1984). Selection for other agronomically important traits such as disease resistance, plant architecture, grain colour and type, broad environmental adaptation and time to maturity were also included so that materials produced would be commercially acceptable.

Since selection criteria for increased N_2 fixation that were used to produce the RIZ lines were indirect; evaluation of the effect of this selection programme was made by measuring the amount of N_2 fixed and proportion of N derived from fixation in selected and parental lines. Isotope dilution methods were first used by Rennie and Kemp (1983, 1984) and Ruschel et al. (1982) to quantify N_2 fixation in beans and to investigate differences between genotypes. Witty (1983) noted some problems in the use of ^{15}N-labelling studies for measuring N_2 fixation in the field, however these can be largely overcome if the ^{15}N is added with a carbon source in order to stabilize the enrichment of plant available soil N and reduce errors associated with decreasing soil enrichment (Giller and Witty, 1987).

Many characteristics determined by the plant line may contribute to N_2 fixation capacity. We have classified these plant attributes into five main group (Table 1). While a simple scheme of classification obviously ignores the complex interactions between each group of characteristics and the environment, it can provide a useful framework for consideration of such interactions. In the work described, experiments were restricted to lines with a similar growth habit and time to maturity in order to simplify our comparisons; other experiments have compared bean cultivars of widely differing duration which could account for much of the variation in N_2 fixation (Graham and Rosas, 1977; Rennie and Kemp, 1983).

Estimates of N_2 fixation by some parental lines were compared with those of several lines (RIZ lines) produced from a breeding programme for enhanced nitrogen fixation at CIAT (CIAT, 1983). The parental lines had been included in the programme for characters such as disease resistance and good architecture, as well as for abundant nodulation. All the lines have agronomic promise, possessing resistance to several diseases, good architecture, grain type and high yield. N-uptake curves, N partitioning and nodulation patterns were studied in a separate experiment to allow comparison of N_2 fixation with indirect selection criteria, in the hope that improvements in the ongoing breeding programme could be made.

Table 1. Characteristics of the plant genotype which may be important in determining the amount of N_2 fixed by the symbiosis

Character	Factors
Duration	Time to maturity, time to flowering, delayed leaf senescence.
Vigor	Plant size, photosynthetic area, plant architecture, N sinks.
Nodulation	Number, size and longevity of nodules, host/Rhizobium strain specificity for nodulation, nodulation under high soil nitrogen.
Physiology of N_2 fixation	Efficiency of carbon use by nitrogenase, specific nitrogenase activity of nodules, carbon and nitrogen transport to and from nodules, oxygen transport mechanisms.
Fertilizer uptake	Ability to exploit or ignore soil and fertilizer nitrogen.

Materials and methods

Four promising breeding lines (RIZ27, RIZ29, RIZ30 and RIZ36) were compared with five parental lines of similar seed types (BAT76, BAT332, BAT1297, BAT1554 and A268). The RIZ lines, produced from crossing and recurrent selection, have complex pedigrees (CIAT 1984). The nine indeterminate small-seeded bush beans have similar growth habit and duration and all the seed for these experiments was from a crop grown at the CIAT Palmira station.

Two experiments were planted in adjacent plots in a field at the CIAT research centre, Palmira (isohyperthermic Aquic Hapludoll). A cover crop of maize was grown before the experiments and the aerial parts removed to deplete the plant-available N in the soil. The experiments were planted in October 1984, with rows 50 cm apart and seeds spaced at 10 cm. A dwarf sorghum cultivar (SC 326-6), used to estimate the pool of available soil N, was sown at a rate of 0.8 g per m row. Granular inoculum containing $>10^9$ cells g^{-1} of *Rhizobium leguminosarum* bv. *phaseoli* strains CIAT 144, CIAT 166 and CIAT 899 in equal numbers, was added to all rows at sowing ($1 g m^{-1}$). Insecticides and fungicides were applied as needed throughout the season.

Experiment 1

The ten treatment plots (2 m × 1 m) were randomized within six replicate blocks (10 m × 2 m). Isotope-labelled fertilizer, $10 kg N ha^{-1}$ of $(NH_4)_2SO_4$ (29.3 atom % ^{15}N excess), was applied in solution with sucrose 7 days prior to sowing. The amount of sucrose was calculated to give a C:N ratio of 10:1 in the solution. The entire 120 m^2 area was labelled, but the fertilizer was measured and added to each treatment plot individually. The central 0.8 m^2 of each plot was harvested 56 days after sowing and roots, shoots and pods separated. The samples were dried, weighed and ground. Nitrogen content was determined by Kjeldahl digestion and an automated indophenol-blue method (Novozamsky et al., 1974). Nitrogen in the digests was concentrated by a Conway microdiffusion technique (Conway, 1939) and ^{15}N enrichments were measured using a Micromass 622 mass spectrometer. Fertilizer recovery and N_2 fixation were calculated using the equations of Witty (1983), with correction for seed N in some cases (Witty and Ritz, 1984).

Experiment 2

Ten treatment plots (7 m × 2.5 m) were randomized within four replicate blocks (70 m × 2.5 m). Unlabelled $(NH_4)_2SO_4$ equivalent to that applied in Experiment 1 was broadcast over the plots. Harvests (1.5 m × 0.5 m) were taken at 13 days and subsequently at weekly intervals until the plants were 56 days old. Root systems were excavated carefully to recover the nodules. Nodules were separated, counted, dried and weighed. Roots, shoots and pods were dried, weighed and analyzed separately for N content. A final harvest (7 m^2) was taken at 93 days and yield determined.

Data were analyzed by one-way analysis of variance procedures, and when the F-test was significant means were compared at the 0.05 probability level using Duncan's multiple range test.

Results

Despite heavy rainfall at Palmira shortly after planting, the plants established and grew well to give a good crop dry weight at 56 days (Table 2) and grain yield at maturity. Estimates of N_2 fixed in the crop were carried out during pod-fill before leaf senescence and fall. In most lines at 56 days the pods accounted for approximately 30% of shoot dry matter and 30 to 40% of the N was in the pods. For certain lines, such as RIZ27, however the pods constituted only 22% of shoot dry matter and 28% of shoot N, while

Table 2. Dry weight and nitrogen content ($kg ha^{-1}$) of component parts of beans harvested at 56 days. (Experiment 1)

Line	Stem and leaves		Pods		Roots and nodules	
	Dry wt.	N	Dry wt.	N	Dry wt.	N
A 268	1820 a	43 ab	970 ab	33 a	180 bc	2.4 c
RIZ 30	1650 ab	42 ab	960 ab	32 a	180 bc	2.4 c
RIZ 36	1780 a	44 ab	810 bcd	28 abc	230 a	3.2 ab
RIZ 27	1720 ab	46 a	490 ef	18 d	160 cd	2.4 c
RIZ 29	1910 a	48 a	880 bc	30 ab	230 a	3.3 a
BAT 76	1840 a	44 ab	690 cd	23 cd	230 a	2.9 bc
BAT 1297	1300 c	32 c	1140 a	34 a	140 d	2.0 d
BAT 332	1330 c	37 bc	670 de	24 bcd	170 bc	2.4 c
BAT 1554	1440 bc	37 bc	450 f	16 d	200 ab	2.7 c

Means having a common letter are not significantly different at $p < 0.05$ by Duncan's multiple range test.

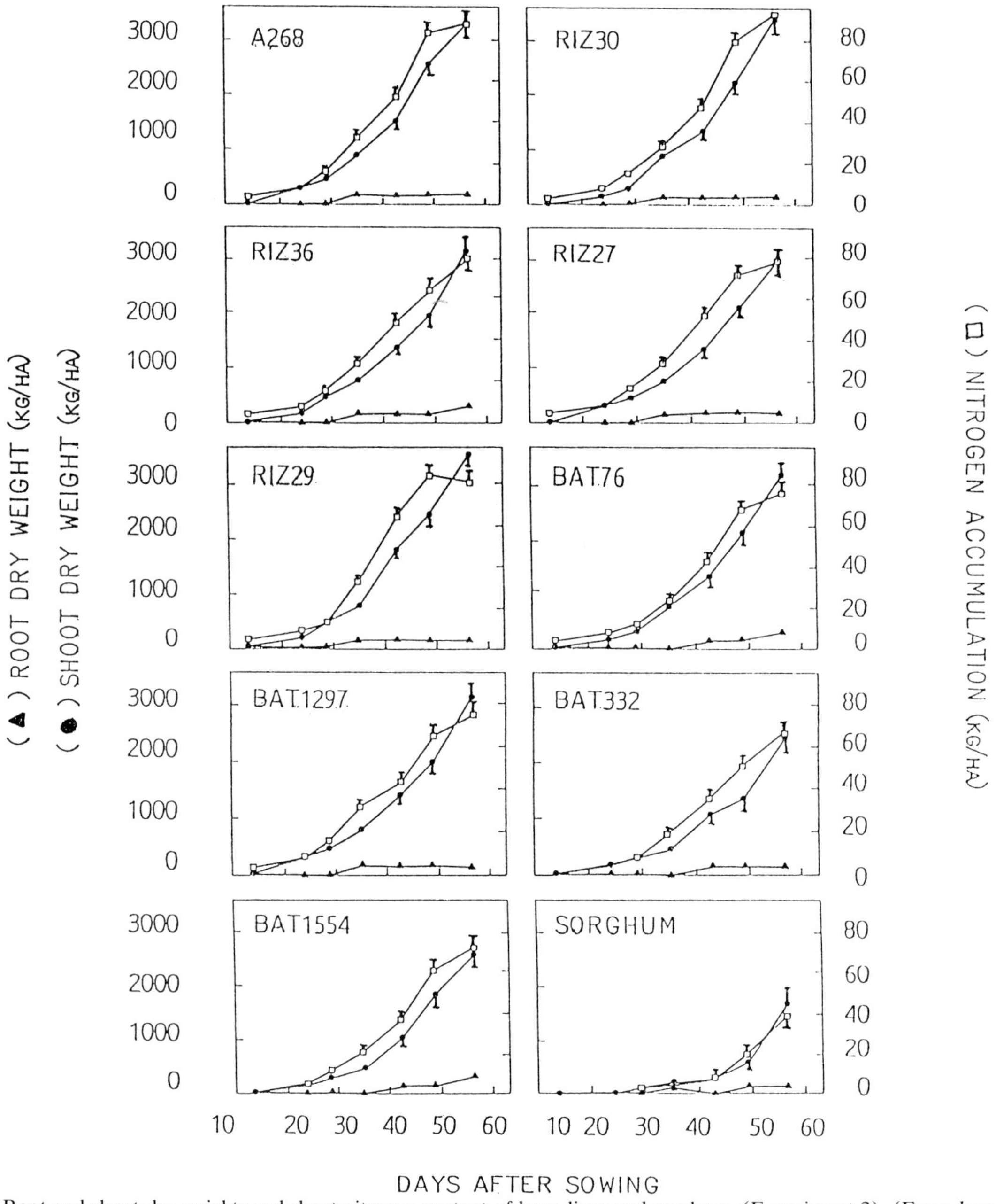

Fig. 1. Root and shoot dry weights and shoot nitrogen content of bean lines and sorghum. (Experiment 2). (Error *bars* indicate the standard error of the treatment means at each harvest).

the pods of BAT1297 contained 46% of shoot dry matter and 52% of shoot N.

Significant differences were found among lines in total shoot N. BAT1554 had 53 kg N ha^{-1}, only 69% of the shoot N of RIZ29 which had 77 kg N ha^{-1} (Table 3). The roots contained a small proportion of the total dry weight (<10%; Table 2) and very small amounts of N (maximum 3.3 kg ha^{-1}). RIZ29 and RIZ36 had the highest root N content and were significantly different from the other lines. Fertilizer recovery was low in all bean lines, ranging from 7.5% in BAT1554 to 10.6% in RIZ29. Sorghum had a smaller N content and fertilizer recovery than the beans.

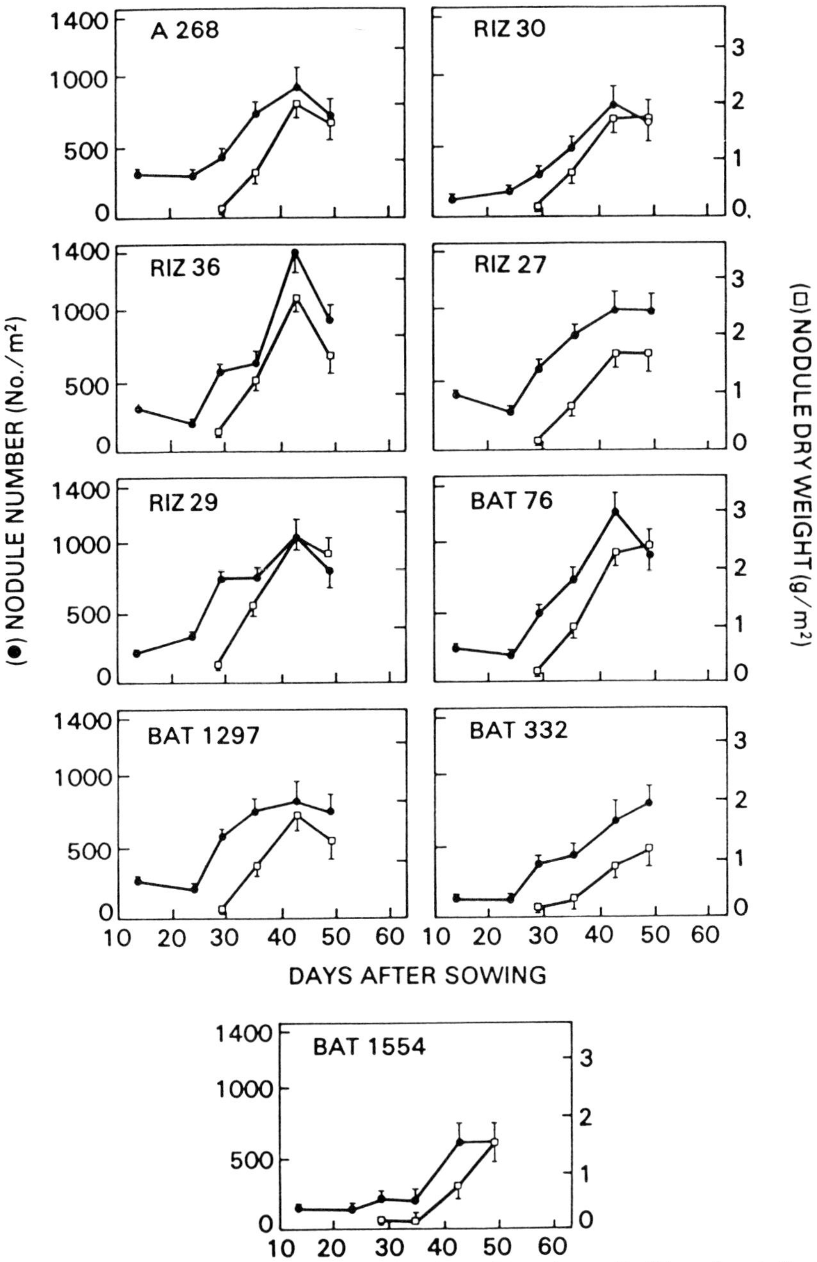

Fig. 2. Number of nodules and nodule dry weight on roots of bean lines. (Experiment 2). (Error *bars* indicate the standard error of the treatment means at each harvest).

Parental line A268 fixed the greatest amount (36 kg N ha^{-1}) and highest proportion (47%) of shoot N (Table 3). The differences among RIZ lines and most parental lines were not significant. BAT1554 had the smallest amount of fixed N_2 (18 kg N ha^{-1}) and also the smallest proportion of N from fixation (32%). The line ranking for amount of fixed N was not the same as that for the percentage of shoot-N from fixation. The N content in the planted seeds of the lines varied slightly, but when the N from fixation was calculated to correct for seed N (Witty and Ritz, 1984) in both sorghum and beans there was little effect on the amount of N_2 fixed and no change in line ranking with 34 kg N ha^{-1} for A268 to 17 kg N ha^{-1} for BAT1554.

In Experiment 2 maximum rates of accumulation of dry matter and shoot N were reached at a similar time in all lines (between 30 and 48 days). Up to 56 days dry matter was still increasing in all lines, whereas at that time the rate of N accumulation was in most cases considerably reduced (Fig. 1). Sorghum established slowly but grew quickly after 48 days absorbing most of its nitrogen between 40 days after sowing and the final harvest.

Significantly more root nodules ($>290\ m^{-2} = 15\ plant^{-1}$) were formed by 13 days after sowing in A268, RIZ36 and RIZ27 than in other lines (Fig. 2). In most lines both the number (800 to 1280 m^{-2}) and weight (1.6 to 2.8 g m^{-2}) of nodules reached maximum values 41 days after sowing. However, nodule number and nodule mass of BAT332 and BAT1554 increased later with the highest values for nodule weight recorded at the 48 day harvest (BAT1554 = 1.47 g m^{-2}; BAT332 = 1.27 g m^{-2}). Nodules had senesced by the 56 day harvest.

Discussion

Measurements of N_2 fixation

A substantial amount of N was derived from fixation in all of the bean lines when grown in a fertile soil at Palmira. Although the RIZ lines tended to fix more N than the BAT lines, the differences were not significant, and the line that fixed most N, A268, was a parental line. This result suggests that the breeding programme was not effective in increasing fixation levels over the previously identified good-fixing lines for this growth habit and maturity group (i.e. BAT 76, BAT 332 and A268). In contrast, improvement of very poor-fixing cultivars to levels already available in the crop, has been more successful (CIAT, 1990; St. Clair and Bliss, 1991; St. Clair et al. 1988).

Recovery of fertilizer was small in all cases (<11%) because N salts were added with a readily available carbon source to immobilize the labelled isotope. The amount of N derived from fertilizer was less than 1 kg N ha^{-1} in all cases, but more than 50% of plant N was derived from the soil because the soil N pool was large. Addition of fertilizer in this form resulted in similar estimates of N_2 fixation irrespective of the control plant employed (Giller and Witty, 1987). The sorghum reference crop established slowly as it was sown deeply and this resulted in delayed uptake of nitrogen. If a gradual decline in the ^{15}N enrichment of the plant-available soil N had occurred, a greater proportion of the N absorbed by sorghum would have been at a lower enrichment causing an underestimation of N_2 fixation in the beans (cf Witty, 1983). Root N constituted only a small fraction (5%) of the total in all lines and although complete recovery of roots and nodules was unlikely, neither this nor the amount of N from the sown seed would have had a substantial effect on the calculations of N_2 fixed. However, it is recognized that the amounts of fixed N_2 lost to the soil may have varied between lines. The amounts of N_2 fixation are comparable to those reported in similar bean cultivars in Brazil where up to 32 kg N ha^{-1} from fixation was found (Duque et al. 1985). By contrast, Rennie and Kemp (1983) reported generally greater amounts of N_2 fixation (40 to 110 kg N ha^{-1}) in longer duration beans grown in temperate regions.

Nitrogen uptake and nodulation

Total shoot N did not give the same ranking as either the amount or proportion of N_2 fixed, or the fertilizer recovery (which can serve as a measure of soil N uptake ability). This suggests that there is genetic variability in both the ability to fix N and to absorb N from the soil. At the 56-day harvest, all lines showed chlorosis associated with maturation and pod-fill and the majority of nodules had senesced, suggesting that further shoot growth was due to some extent to remobilization of N in the plants. Differences in partitioning of N to the pods were particularly striking, ranging from 52% of shoot N in BAT1297 to only 28% of shoot N in RIZ27 (Table 2), even though the total N in these lines was similar.

In general, lines which fixed large amounts of N (e.g. A268, RIZ36) had more nodules and greater nodule weight than the less effective lines (e.g. BAT332 and BAT1554) but this was not true in all cases (cf. RIZ30 and BAT1297). Differences were observed in the patterns of nodule development, in particular the two lines with the smallest amounts of N_2 fixation (BAT332, BAT1554) formed nodules later than the other lines (Fig. 2).

Plant characters related to the amount of N_2 fixation

With the information available from these field experiments, it is interesting to assess the importance of different plant characteristics (Table 1) in determining the amounts of N_2 fixation. All lines were of similar growth duration, and there was no evidence that fixation continued later into pod-fill in some lines since nodules were senescent in all lines by the 56 day harvest. Plant vigor is presumed to be a character with which the amount of N_2 fixation is related since plants with greater shoot weight have larger photosynthetic area to provide carbon for N_2 fixation (Table 2). However these results suggest that there are potential differences in N_2 fixation (from 38–47% N from fixation) between plants with similar total N (77–75 kg N ha^{-1}; cf. A268 and RIZ 29, Table 3). Some variation in exploitation of soil N was indicated by the variation in fertilizer recoveries between lines; this was smaller than the variation in amounts of N_2 fixation (Table 3), but together with differences in concentration of N may account for poor correlations between vigor and fixation.

Differences both in nodule weight and, perhaps more strikingly, in the time required to form nodules appeared to be important in determining the amounts of N_2-fixed. However, good fixation by lines such as RIZ30 indicate the need for additional studies on specific nodule activity and carbohydrate use efficiency.

Improvements in selection criteria

Selection based on growth and nodulation in sand culture in the glasshouse was abandoned when it was discovered that materials had poor agronomic characters when grown in the field (CIAT 1982). Care has since been taken to include multiple disease resistance so that selected materials have greater agronomic potential. In addition to selection for yield under low N conditions, early nodulation characters have been incorporated into the selection programme due to their indicated importance in the present study. Obviously vigor and the amount of nodulation are also integrally related to the amount of N_2 fixed by a plant and selection for extended N_2 fixation into the period of pod-fill (late nodule senescence) may be advantageous. As these plant traits are not all combined in the best-fixing lines, the potential to increase fixation in the species through appropriate choice of parents and selection criteria is indicated.

Table 3. Nitrogen fixation in bean lines harvested at 56 days. (Experiment 1)

Line	Shoot N (ka ha^{-1})	% Fertilizer recovery	Mean atom % ^{15}N excess	N-fixed (kg ha^{-1})	%N from fixation
A 268	75 a	8.7 ab	0.338 b	36 a	47 a
RIZ 30	73 ab	8.8 ab	0.352 b	33 ab	45 ab
RIZ 36	71 ab	9.0 ab	0.367 b	30 abc	43 ab
RIZ 27	64 abc	7.6 b	0.346 b	29 abc	46 ab
RIZ 29	77 a	10.6 a	0.398 b	29 abc	38 ab
BAT 76	67 ab	8.8 ab	0.395 b	27 abc	40 ab
BAT 1297	65 abc	8.5 ab	0.385 b	27 abc	40 ab
BAT 332	61 bc	8.7 ab	0.418 b	21 bc	35 ab
BAT 1554	53 c	7.5 b	0.434 b	18 c	32 b
Sorghum	28 d	6.5 b	0.648 a		

Means having a common letter are not significantly different at $p < 0.05$ by Duncan's multiple range test.

Acknowledgements

Joint contribution Bean Microbiology, Centro Internacional de Agricultura Tropical (CIAT), Cali, Colombia and Soils Division, Rotamsted Experimental Station, U.K. The Rothamsted collaboration was funded through Overseas Development Administration Research Scheme R3648A.

We thank Dr. Steve Temple for help in choosing the bean lines for this study, the staff of the Bean Microbiology section at CIAT for assistance in carrying out these experiments, and Moyra Smith and Phil Edwards for the chemical analyses.

References

Barnes D K, Heichel G H, Vance C P and Ellis W R 1984 A multiple-trait breeding programme for improving the symbiosis for N_2 fixation between *Medicago sativa* L. and *Rhizobium meliloti*. Plant and Soil 82, 303–314.

CIAT 1978 Annual Report of the Bean Production Programme 75p.

CIAT 1982 Annual Report of the Bean Production Programme. 234p.

CIAT 1983 Annual Report of the Bean Production Programme. 238p.

CIAT 1984 Annual Report of the Bean Production Programme. 311p.

CIAT 1990 Annual Report of the Bean Programme. 350p.

Conway E J 1939 Microdiffusion Analysis and Volumetric Error. Crosby-Lockwood, London.

Duque F F, Neves M C P, Franco A A, Victoria R L and Boddey R M 1985 The response of field grown *Phaseolus vulgaris* to *Rhizobium* inoculation and the quantification of N_2 fixation using ^{15}N. Plant and Soil 88, 333–343.

Giler K E and Witty J F 1987 Slow release fertilizers can improve the accuracy of field estimates of N_2 fixation using ^{15}N. Soil Biol. Biochem. 19, 459–463.

Graham P H 1981 Some problems of nodulation and symbiotic nitrogen fixation in *Phaseolus vulgaris* L.: A review. Field Crops Res. 4, 93–112.

Graham P H and Rosa J C 1977 Growth and development of indeterminate bush and climbing cultivars of *Phaseolus vulgaris* L. inoculated with *Rhizobium* J. Agric. Sci. Camb. 88, 503–508.

Graham P H and Temple S R 1984 Selection for improved nitrogen fixation in *Glycine max* (L.) Merr. and *Phaseolus vulgaris* L. Plant and Soil 82, 315–327.

Greder R R 1983 Heritabilities and associations of nodule mass and recovery of *Bradyrhizobium japonicum* strain 110 in soybeans. M. Sc. Thesis, University of Minnesota, St. Paul 43 p.

Kueneman E A, Root W R, Dashiell K E and Hohenberg J 1984 Breeding soybeans for the tropics capable of nodulating effectively with indigenous *Rhizobium* spp. Plant and Soil 82, 387–396.

McFerson J R 1983 Genetic breeding studies of dinitrogen fixation in common bean, *Phaseolus vulgaris* L. Ph.D. Thesis, University of Wisconson. 147 p.

McFerson J R, Bliss F A and Rosas J C 1982 Selection for enhanced nitrogen fixation in common beans, *Phaseolus vulgaris*. *In* Biological Nitrogen Fixation Technology for Tropical Agriculture. Eds. P H Graham and S C Harris. pp 39–44. CIAT, Cali.

Miller J C, Zary K W and Fernandez G C J 1986 Inheritance of N_2 fixation efficiency in cowpea. Euphytica 35, 551–560.

Novozamsky I, van Eck R, van Schouwenburg J C and Walinga I 1974 Total nitrogen determination in plant material by means of the indophenol-blue method. Neth. J. Agric. Sci. 22, 3–5.

Nutman P S 1984 Improving nitrogen fixation in legumes by plant breeding: The relevance of host selection experiments in red clover (*Trifolium pratense* L.) and subterranean clover (T. *subterraneum* L.). Plant and Soil 82, 285–301.

Rennie R J and Kemp G A 1983 N_2 fixation in field beans quantified by ^{15}N isotope dilution. II. Effect of cultivars of beans. Agron. J. 75, 645–649.

Rennie R J and Kemp G A 1984 ^{15}N-determined time course for N_2 fixation in two cultivars of field bean. Agron. J. 76, 146–154.

Ruschel A P, Vose P B, Matsui E, Victoria R L and Tsai Saito S M 1982 Field evaluation of N_2 fixation and N-utilization by *Phaseolus* bean varieties determined by ^{15}N isotope dilution. Plant and Soil 65, 397–407.

St. Clair D A and Bliss F A 1991 Intrapopulation recombination for ^{15}N-determined dinitrogen fixation ability in common bean. Plant Breeding 106, 215–225.

St. Clair D A, Wolyn D J, DuBois J, Burris R H and Bliss F A 1988 A field comparison of N_2 fixation determined with ^{15}N-depleted ammonium sulfate and ^{15}N-enriched ammonium sulfate in selected inbred backcross lines of common bean. Crop Sci. 28, 773–778.

Witty J F 1983 Estimating N_2 fixation in the field using ^{15}N labelled fertilizer: Some problems and solutions. Soil Biol. Biochem. 15, 631–699.

Witty J F and Ritz K 1984 Slow release ^{15}N fertilizer formulations to measure N_2 fixation by isotope dilution. Soil Biol. Biochem. 16, 657–661.

Plant and Soil **152**: 115–121, 1993.

PLSO BNF12

Time course of N_2 fixation in common bean (*Phaseolus vulgaris* L.)

J.J. PEÑA-CABRIALES[1], O.A. GRAGEDA-CABRERA[1], V. KOLA[2,3] and G. HARDARSON[2]
[1]*CINVESTAV, Irapuato, México*; [2]*FAO/IAEA Laboratory, A-2444, Seibersdorf, Austria*;
[3]*Present address: Instituti Kerkimeve Bujqesore, Lushnje, Tirana, Albania*

Key words: common bean, nitrogen assimilation, nitrogen fixation, nitrogen partitioning

Abstract

Field and greenhouse experiments were conducted to assess the nitrogen fixation rates of four cultivars of common bean (*Phaseolus vulgaris* L.) at different growth stages. The ^{15}N isotope dilution technique was used to quantify biological nitrogen fixation. In the greenhouse, cultivars M4403 and Kallmet accumulated 301 and 189 mg N plant^{-1}, respectively, up to 63 days after planting (DAP) of which 57 and 43% was derived from atmosphere. Under field conditions, cultivars Bayocel and Flor de Mayo RMC accumulated in 77 DAP, 147 and 135 kg N ha^{-1}, respectively, of which approximately one-half was derived from the atmosphere. The rates of N_2 fixation determined at different growth stages increased as the plants developed, and reached a maximum during the reproductive stage both under field and greenhouse conditions. Differences in translocation of N were observed between the cultivars tested, particularly under field conditions. Thus, the fixed N harvest index was 93 and 60 for cultivars Flor de Mayo and Bayocel, respectively. In early stages of growth, the total content of ureides in the plants correlated with the N fixation rates. The findings reported in the present paper can be used to build a strategy for enhancing biological N_2 fixation in common bean.

Introduction

It has been widely stated that the common bean is a poor N_2 fixing plant (Graham, 1981; Rosas, 1983). Two aspects have been considered of major importance in determining N_2 fixation effectiveness: (i) environmental factors (Saito, et al., 1984; Smith et al., 1985, 1988) and (ii) the plant genotype (Hardarson et al., 1993).

Time course of N_2 fixing activity under field conditions has been studied for soybean (Zapata et al., 1987a) and fababean (Zapata et al., 1987b). Soybean were found to have limited initial fixation, and it was only at R_1 stage of growth (74 DAP) when substantial N_2 fixation (3 kg N ha^{-1}day^{-1}) occurred. This high N_2 fixation was maintained for only approximately 20 days until R_5. In the final harvest approximately 50% of N in the soybean crop or 100 kg N ha^{-1} was derived from the atmosphere. Therefore, the possibility exists for enhancing nitrogen fixation in soybean during vegetative growth. This could be done by improving inoculation methods or breeding for early nodulation.

Fababean (Zapata et al., 1987b) had a very different time course of nitrogen fixation. In that crop most of the nitrogen assimilation occurred during the period 60 to 110 DAP, with approximately 75% Ndfa (N derived from atmosphere) and 160 kg N ha^{-1} fixed. With fababean being so effective, it will be difficult to enhance fixation. Few such studies have been performed for common bean. The objective of the present study was to investigate the time course of N_2 fixation in common bean under greenhouse and field conditions and to identify growth periods where

N_2 fixation could be enhanced. The study was performed using the ^{15}N isotope dilution technique.

Materials and methods

Greenhouse experiment

The first experiment was performed in the greenhouse of the FAO/IAEA laboratory Seibersdorf, Austria. Two cultivars of *Phaseolus vulgaris* L.cv. Kallmet (type I) and M4403 (type II) were grown in 3 kg pots containing Typic Eutrocrepts soil (Zapata et al., 1987a) from the experimental field at Seibersdorf. Non-nodulating soybean (*Glycine max* L. Merrill) cv. Chippewa were used as a reference crop. The experiment was conducted in a randomized block design with four replicates and each pot containing two plants. 10 ppm N of 10.76% ^{15}N atom excess ammonium sulphate were applied as a solution five days after planting. The pots were watered when necessary. The plants were harvested 28, 35, 42, 49, 63 and 77 days after planting (DAP). The harvested material was oven dried at 72°C. Total N determinations were done by Kjeldhal method (Eastin 1978) and % ^{15}N atom excess by mass spectrometer (Fieldler and Proksh, 1975). The amount of N_2 fixed by bean was determined using the isotope dilution technique (McAuliffe et al., 1958; Fried and Middelboe, 1977).

Field experiment

This experiment was performed at the experimental field of INIFAP, Guanajuato, Mexico. The soil is a typic Pellusterts with a pH of 7.7, organic matter content of 1.2% and clay in texture. Two common bean cultivars, Flor de Mayo RMC and Bayocel, were studied. Both cultivars had been characterized previously as efficient in N_2 fixation. Maize (*Zea mays* L.) and Sorghum (*Sorghum vulgare*) were included as non-fixing reference crops. The experiment included 4 crop treatments, 8 sampling times and 6 replicates laid out in a randomized block design. Each crop was grown in 5.5 m rows with 40 cm interrow spacing. 10 kg N ha^{-1} of 10.075% ^{15}N atom excess ammonium sulphate was applied as a solution to 1.5 m of the above rows. The beans were inoculated at sowing with a multistrain inoculant prepared by the Microbial Ecology Laboratory of CINVESTAV using sterile peat as carrier. The whole experimental area was amended with $CaHP_2O_4$ at a rate of 40 kg ha^{-1} three days before sowing.

Whole plant samples (3 plants per treatment and replicate) were taken at 42, 49, 56, 63, 70, 77, 84 and 91 DAP. The following parameters were measured: dry weight of the different organs; leaf area using a portable area meter (Li-Cor Model LI-3,000); acetylene reduction activity (ARA) according to the method of Dart (1972) using a gas chromatograph (Hewlett-Packard series 5880) with an ionization flame detector and a Porapack column; ureides production determined in dry matter of various organs using the method of Patterson (1982). One meter of each of the labelled rows was harvested, weighed and separated into pods and straw. The samples were chopped into 1 to 2 cm pieces, subsampled and oven dried at 70°C. The dried samples were ground to pass through a 0.5 mm sieve. Total N and ^{15}N isotope ratio were determined by 1500 ANA N analyser coupled to a VG mass spectrometer at the IAEA Laboratory, Seibersdorf. Calculation of nitrogen fixation was performed as described for the previous experiment. Statistical analysis was done using the least significant different multiple-range-test.

Results

The accumulation of total and fixed nitrogen throughout the growth cycle of the four cultivars studied is presented in Figure 1. Similar patterns of accumulated and fixed N_2 were observed for plants grown in either the field or the greenhouse. In both cases, the maximum accumulation of N occurred at the stage of physiological maturity. From early flowering to physiological maturity, the plants accumulated more than 50% of the total dry matter.

The cultivars studied in the greenhouse started showing differences with respect to accumulated fixed N at between 49 and 63 DAP. The cultivar MM4403 accumulated more nitrogen than the

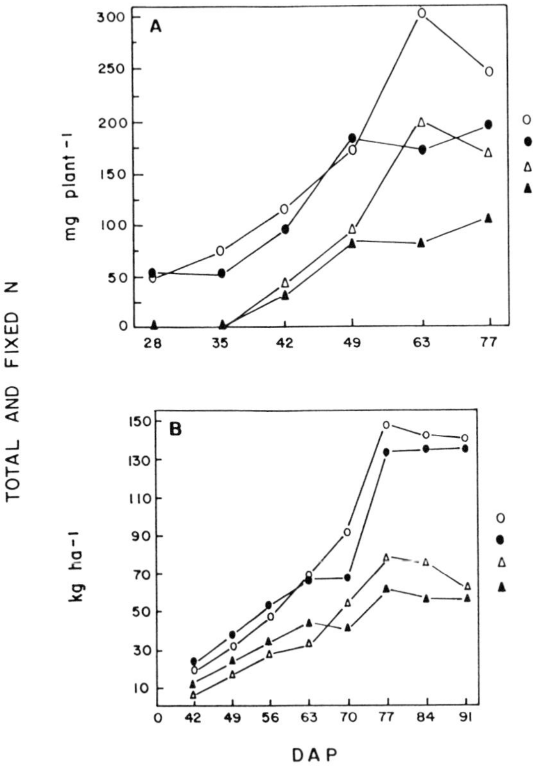

Fig. 1. Total nitrogen and nitrogen derived from the atmosphere accumulated throughout the growth cycle by four cultivars of common bean (*Phaseolus vulgaris*) grown in the greenhouse (**A**) and field conditions (**B**).

A ○ c.v. M4403-total N
● c.v. Kall.-total N
△ c.v. M4403-fixed N_2
▲ c.v. Kall.-fixed N_2

B ○ c.v. Bayocel-total N
● c.v. Flor de Mayo RMC-total N
△ c.v. Bayocel-fixed N_2
▲ c.v. Flor de Mayo RMC-fixed N_2

cultivar Kallmet. Similar behaviour was observed for the cultivars grown in the field. Under those conditions however, the differences in total accumulated nitrogen and fixed nitrogen were not as large as those in the greenhouse.

In Fig. 2, the rates of N assimilation of fertilizer, soil, and fixed nitrogen are shown for greenhouse and for field studies. In the greenhouse the rates of nitrogen assimilation increased as the plants grew and they achieved maximum N uptake after flowering. Similarly, in the field studies the cultivars tested showed the highest N_2 fixation and N assimilation rates during the reproductive stage. During maturation no further N assimilation occurred.

The translocation of nitrogen to different plant tissues by the four cultivars studied is shown in Figure 3. Appreciable differences in the nitrogen partitioning were observed in both sets of conditions. In the greenhouse for instance, the two cultivars showed the same amount of nitrogen left in the straw, yet cultivar M4403 translocated a larger proportion of its total nitrogen to pods. Under field conditions, the cultivars translocated a larger proportion of nitrogen to pods compared to the plants growing in the greenhouse. Significant differences were observed in the N translocation pattern in the cultivars tested. For instance, it was interesting to observe that N translocation into pods occurred earlier in cultivar Flor de Mayo than in Bayocel.

In Table 1 the indexes of total nitrogen harvest and the fixed nitrogen are shown with higher values being obtained in the field studies. Under these conditions, significant differences between cultivars were observed. Flor de Mayo translocated a higher proportion of the fixed nitrogen to pods than cv. Bayocel. On the contrary, no significant differences were observed between the cultivars studied in the greenhouse.

The results obtained on other parameters monitored during the growth cycle of the cultivars grown in the field are summarized in Table 2. These parameters include nodulation (nodule dry weight), nitrogenase activity (acetylene reduction), and foliar characteristics (leaf dry weight, leaf area). Nodulation of the two cultivars was only different towards the end of the growing cycle (84 and 97 DAP). At these stages cultivar Bayocel maintained its nodules whereas Flor de Mayo already presented nodule senescence. The acetylene reduction activity showed the same pattern for both cultivars. The activity increased until the plants were at the reproductive stages and then decreased to an undetectable level. Significant differences in acetylene reduction were observed between the cultivars only after the flowering stages. Bayocel maintained a high activity for a longer period of time.

Total leaf dry weight was practically the same for both cultivars for the early growth. However, significant differences in leaf area were observed after 69 DAP when cultivar Bayocel had larger leaf area than Flor de Mayo.

The ureide content of the different organs in the two cultivars studied were particularly abun-

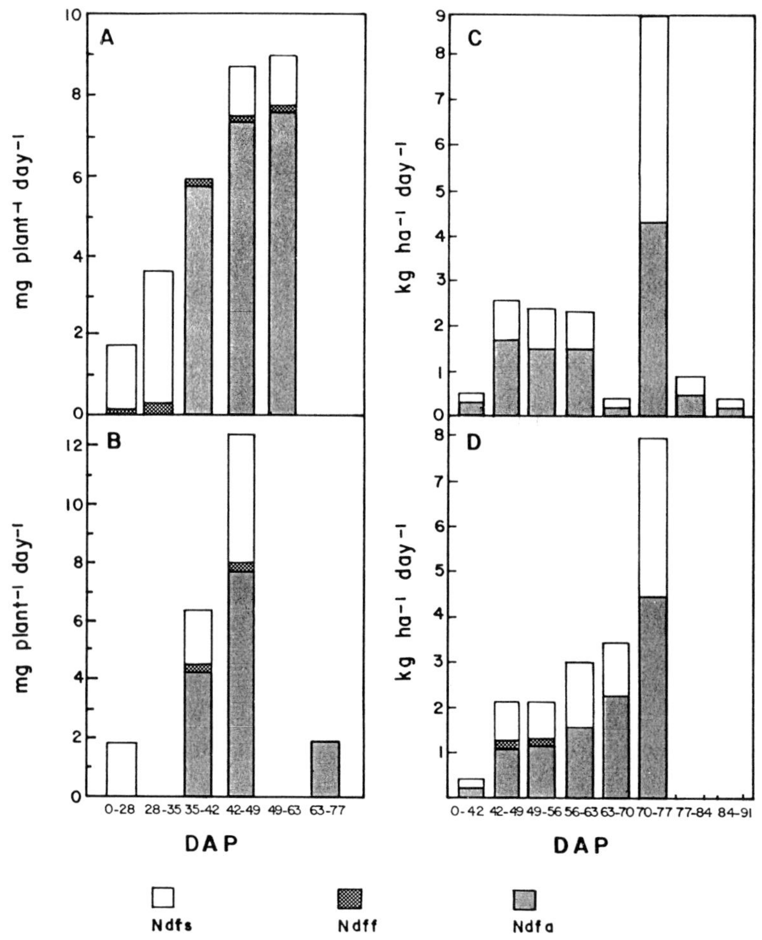

Fig. 2. N assimilation rates of fertilizer (Ndff), soil (Ndfs), and N_2 fixed (Ndfa) by four cultivars of common bean (*Phaseolus vulgaris*) at various intervals, and grown in the greenhouse (**A** and **B**) and field conditions (**C** and **D**)
A Cultivar M4403 C Cultivar Flor de Mayo RMC
B Cultivar Kallmet D Cultivar Bayocel

dant in the stems (Table 3). It was only in this organ that significant differences between cultivars were detected in early stages, i.e. 54 DAP. The cultivar Bayocel showed a consistently higher total ureide concentration throughout the growth cycle of the plant.

Discussion

The present study permits a comparison of the behaviour of different cultivars of common bean grown in the greenhouse and the field. In both cases the cultivars were selected according to their previously known N_2 fixation performance. The N assimilation patterns were influenced by the cultivar. Thus, cultivars Kallmet and Flor de Mayo apparently did not absorb more N during maturity whereas mobilization of N within the plant took place during this period of time.

The highest N assimilation rate for greenhouse and field grown plants occurred during the reproductive stage. Nodule dry weight and acetylene reduction activity had decreased considerably in the field experiment. At that time nodules on lateral roots may be more important in terms of N_2 fixation (Hardarson et al., 1989). Similar results have been reported by Wolyn et al. (1989) who suggest that in common bean nodulation of lateral roots in late stages plays an important role in providing N to the plants. For plants grown in the field, nodulation in lateral

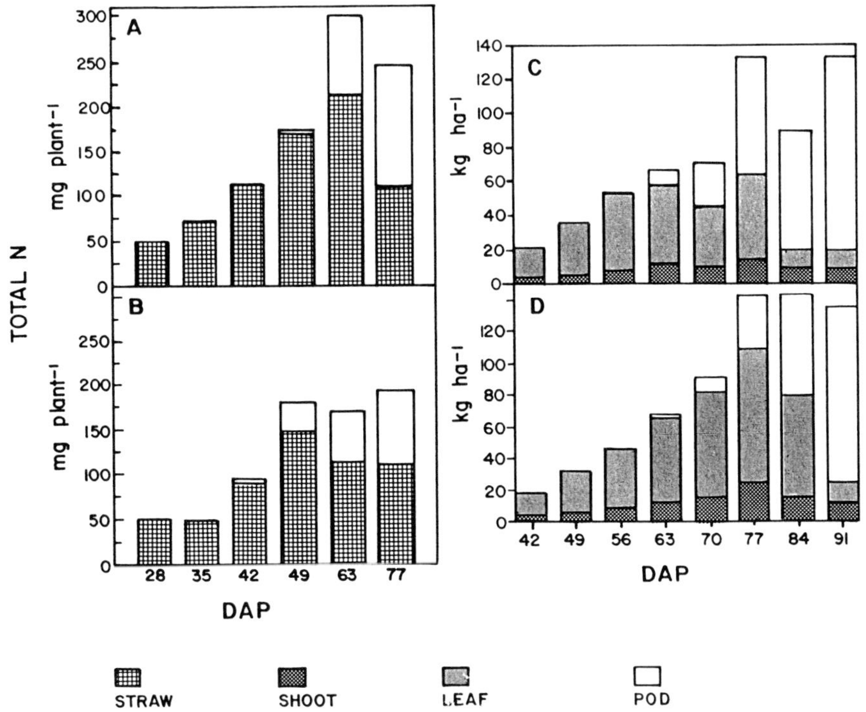

Fig. 3. Nitrogen translocation during the growth cycle of four cultivars of common bean (*Phaseolus vulgaris*) grown in the greenhouse (**A** and **B**) and field conditions (**C** and **D**).
A Cultivar M4403 C Cultivar Flor de Mayo RMC
B Cultivar Kallmet D Cultivar Bayocel

Table 1. Total nitrogen harvest index and fixed nitrogen index in cultivars of common bean (*Phaseolus vulgaris* L.) grown in the field and greenhouse

	Greenhouse		Field	
	c.v. M4403	c.v. Kallmet	c.v. Flor de Mayo	c.v. Bayocel
Nitrogen Harvest Index	45.86	43.69	87.52**	75.95**
Fixed Nitrogen Index	54.19	54.87	93.55**	60.49**

** significantly different ($p < 0.05$).

roots is very difficult to monitor and nodulation as a whole is often underestimated. ^{15}N experiments showing N assimilation rates of fababean and soybean have also shown the highest nitrogen fixing rates during reproductive stages (Zapata 1987 a and b).

During the early growth stages, N was accumulated in the leaves with most of the N translocated later to the pods. Cultivar Flor de Mayo seems to be better than Bayocel in translocating N to the pods. Translocation and accumulation of N in soybean and cowpea is influenced by the host plant cultivar as well as by the strain of *Rhizobium* (Hungria et al., 1987; Neves et al., 1987).

The biomass fractionation into the different plant organs was a characteristic of the specific cultivar. Thus, cultivars M4403 in greenhouse and Bayocel in the field which showed the higher rates of N_2 fixation also produced higher leaf and

Table 2. Nodule dry weight (NDW), acetylene reduction activity (ARA), leaf dry weight (LDW) and leaf area (LA) of common bean at various growth stages

DAP[a]	NDW	ARA	LDW	LA
	Flor de Mayo RMC			
22	0.01	0.52	0.33	58.62
33	0.05	1.03	0.76	148.84
42	0.11	2.05	1.69	353.29
54	0.18	0.91**	5.55	958.97
69	0.17	1.39*	9.91	1541.28*
84	0.05	0.73*	5.33	1117.66*
97	0.00	0.0**	0.40**	79.93**
	Bayocel			
22	0.01	0.34	0.33	61.85
33	0.03	1.41	0.88	184.88
42	0.09	3.11	1.86	362.42
54	0.22	4.34	4.96	995.71
69	0.22	3.55*	11.71	2342.41*
84	0.11*	2.14	9.47	2342.41*
97	0.34*	0.12**	3.19	567.17**

[a] DAP: days after planting.
* Significantly different ($p < 0.05$).
** Significantly different ($p < 0.01$).

Table 3. Distribution of ureide compounds in tissues of common bean (*Phaseolus vulgaris* L.) at different growth stages. (u*M*/plant)

DAP[a]	Nodules	Roots	Stems	Leaves	Pods	Total
	Flor de Mayo RMC					
22	0.29	1.62	4.01	1.07	–	7.18
33	2.18*	1.81	12.57	9.93	–	26.50
42	4.48	4.88	37.82	2.65	–	45.31
54	5.00	4.84	62.66*	2.40	–	74.97*
69	5.32	7.49	52.67*	0.0	27.56*	93.05*
84	2.91	2.00	25.02*	0.0	0.0	25.46*
97	0.0**	0.14	3.25	0.0	0.0	3.40
	Bayocel					
22	0.15	2.39	3.27	1.91	–	7.73
33	1.00	1.89	13.36	14.70	–	30.96
42	2.68	4.47	34.77	13.77	–	47.37
54	6.07	10.81	139.69*	21.40*	–	178.55
69	7.31	13.66	123.34*	0.0	10.91*	55.23
84	2.91	7.23	51.28*	0.0	19.48*	80.75
97	0.48	1.02	5.98	0.0	0.0	7.48

[a] DAP: days after planting.
* Significantly different ($p < 0.05$).
** Significantly different ($p < 0.01$).

pod dry weights. Similar observations on the fractionation of dry matter in common bean cultivars of different N_2 fixing potential have been made (Rosas, 1983).

There were differences in the ability of the bean cultivars studied to remobilize the N within the plant in both field and greenhouse experiments. This observation is particularly relevant in cases such as the commercial Mexican cultivars tested, which do not show a direct relation between higher % of N derived from the atmosphere and higher translocation of N to the pods.

It should be possible to enhance biological N_2 fixation in common bean by increasing nitrogen fixation during vegetative and early reproductive stages. This may be achieved by improved inoculation methods or by breeding common bean for early nodulation and corresponding fixation.

References

Atkins C A, Pate J S, Ritchie A and Peoples M B 1982 Metabolism and translocation of allontoin in ureide-producing grain legumes. Plant Physiol. 70, 1–6.

Dart P J, Day J M and Harris D 1972 Assay of nitrogenase activity by acetylene reduction. *In* Use of Isotopes for Study of Fertilizer Utilization by Legume Crops, FAO/IAEA, Vienna pp 85–100.

Franco A A, Pereira J C and Neyra C 1979 Seasonal patterns of nitrate reductase and nitrogenase activities in *Phaseolus vulgaris* L. Plant Physiol. 63, 421–424.

Fried M and Middelboe V 1977 Measurement of amount of fixed nitrogen by a legume crop. Plant and Soil 47, 713–715.

Graham P H 1981 Some problems of nodulation and symbiotic nitrogen fixation in *Phaseolus vulgaris* L. Rev. Field Crops Res. 4, 93–112.

Hardarson G, Bliss F A, Cigales-Rivero M R, Henson R A, Kipe-Nolt J A, Longeri L, Manrique A, Peña-Cabriales J J, Pereira P A A, Sanabria C A and Tsai S M 1993 Genotypic variation in biological nitrogen fixation by common bean. Plant and Soil 152, 59–70.

Hardarson G, Golbs M and Danso S K A 1989 Nitrogen fixation in soybean (*Glycine max* L. Merrill) as affected by nodulation patterns. Soil. Biol Biochem. 21, 783-787.

Hungria M and Neves M C P 1987 Cultivar and *Rhizobium* strain effect on nitrogen fixation and transport in *Phaseolus vulgaris* L. Plant and Soil 103, 111–121.

Neves M C P, Didonet A D, Duque F F and Döbereiner J (1985) *Rhizobium* strain effects on nitrogen transport and distribution in soybean. J. Exp. Bot. 169, 1179–1192.

Neves MC and Hungria M 1987 The physiology of nitrogen fixation in tropical grain legumes. Crit. Rev. Plant Sci. 3, 269–321.

Patterson T G, Glenister R and LaRue T A 1982 Simple estimate of ureides in soybean tissue. Anal. Biochem. 119, 90–95.

Rosas J C 1983 Partitioning of dry matter, nitrogen fixation, and seed yield of common bean (*Phaseolus vulgaris* L.) influenced by plant genotype and nitrogen fertilization. Ph. D. Thesis. University Wisconsin, Madison, WI. 121 p.

Rosendahl L 1984 *Rhizobium* strain effects on yield and bleeding sap amino compounds in *Pisum sativum*. Physiol. Plant 60, 215–220.

Saito S M T, Montalheiro M N S, Victoria R L and Reichardt K 1984 The effects of N fertilizer and soil moisture on the nodulation and growth of *Phaseolus vulgaris*. J. Agric. Sci. 103, 87–93.

Shelp B J and Ireland R J 1985 Ureide metabolism in leaves of nitrogen-fixing soybean plants. Plant Physiol. 77, 779–787.

Smith D L, Dijak M and Humme D J 1988 The effect of water deficit on N_2 fixation (C_2H_2) by white bean and soybean. Can. J. Plant. Sci. 68, 957–967.

Smith DL and Humme DJ 1985 Effects of irrigation and fertilizer N on N_2 (C_2H_2) fixation and yield of white bean and soybean. Can. J. Plant Sci. 65, 307–316.

Streeter J G 1979 Allantoin and allantoic acid in tissues and stem from field-grown soybean plants. Plant Physiol. 63, 478–480.

Thomas R J, Feller V and Erismann K H 1979 The effect of different inorganic nitrogen sources and plant age on the composition of bleeding sap of *Phaseolus vulgaris*. New Phytol. 82, 657–669.

Thomas R J, Feller U and Erismann K H 1980 Ureido metabolism in non-nodulated *Phaseolus vulgaris* L.J. Exp. Bot. 121, 409–417.

Thomas R J, McFerson J R, Schrader L E and Bliss F A 1984 Composition of bleeding sap nitrogen from lines of field-grown *Phaseolus vulgaris* L. Plant and Soil 79, 77–88.

Wolyn D J, Attwell J, Ludden P W and Bliss F A 1989 Indirect measures of N_2 fixation in common bean (*Phaseolus vulgaris* L.) under field conditions: the role of lateral root nodules. Plant and Soil 113, 181–187.

Zapata F, Danso S K A, Hardarson G and Fried M 1987a Time course of nitrogen fixation in field-grown soybean using nitrogen-15 methodology. Agron. J. 79, 172–176.

Zapata F, Danso S K A, Hardarson G and Fried M 1987b Nitrogen fixation and translocation in field-grown faba-bean. Agron. J. 79, 505–509.

Plant and Soil **152**: 123–130, 1993.

PLSO BNF13

Response to inoculation and N fertilization for increased yield and biological nitrogen fixation of common bean (*Phaseolus vulgaris* L.)

P.M. DA SILVA, S.M. TSAI and R. BONETTI
Centro de Energia Nuclear na Agricultura, University of São Paulo, P.O. Box 96, Piracicaba CEP 13400-970, São Paulo, Brazil

Key words: common bean, foliar N, nitrogen fixation, N fertilizer, ^{15}N isotope, *Rhizobium*

Abstract

Common bean (*Phaseolus vulgaris* L.) is able to fix 20–60 kg N ha^{-1} under tropical environments in Brazil, but these amounts are inadequate to meet the N requirement for economically attractive seed yields. When the plant is supplemented with N fertilizer, N_2 fixation by *Rhizobium* can be suppressed even at low rates of N. Using the ^{15}N enriched method, two field experiments were conducted to compare the effect of foliar and soil applications of N-urea on N_2 fixation traits and seed yield. All treatments received a similar fertilization including 10 kg N ha^{-1} at sowing. Increasing rates of N (10, 30 and 50 kg N ha^{-1}) were applied for both methods. Foliar application significantly enhanced nodulation, N_2 fixation (acetylene reduction activity) and yield at low N level (10 kg N ha^{-1}). Foliar nitrogen was less suppressive to nodulation, even at higher N levels, than soil N treatments. In the site where established *Rhizobium* was in low numbers, inoculation contributed substantially to increased N_2 fixation traits and yield. Both foliar and soil methods inhibited nodulation at high N rates and did not significantly increase bean yield, when comparing low (10 kg N ha^{-1}) and high (50 kg N ha^{-1}) rates applied after emergence. In both experiments, up to 30 kg N ha^{-1} of biologically fixed N_2 were obtained when low rates of N were applied onto the leaves.

Introduction

Considerable N is required in grain legumes at the beginning of pod fill as the translocation of N from vegetative parts to the pods is intensive. Studies on soybean have shown that this N drain may be high enough to decrease photosynthetic activity (Dunphy and Hanway, 1976), induce premature leaf senescence and reduce root activity (Harper, 1974, Ramon Garcia and Hanway, 1976). Nutrients supplied through leaves may supplement rapidly those transferred from stems and roots, thus avoiding early leave senescence (Frith and Dalling, 1986). Afza et al. (1987) compared soil and foliar N fertilizer at different growth stages of soybean. Nitrogen fixation was not influenced by the two methods (foliar and soil) of application of 40 kg N ha^{-1} to plants if applied during the pod-filling stage. Higher amounts (100 kg N ha^{-1}) reduced the N_2 fixed, but by the judicious application of N fertilizer during the late stages of growth, it was possible to enhance soybean yields without necessarily inhibiting N_2 fixation.

The idea of applying N through leaves to maximize bean yield is not recent (Miyasaka et al., 1963; Muraoka et al., 1978), and still remains controversial due to several technical difficulties and the lack of consistent results (Rosolem et al., 1982, Machado et al., 1982). The most relevant results were found in soybean by Vasilas et al. (1980), with the use of ^{15}N-labelled urea. An increase of 70% in the uptake efficiency of urea was also correlated to a high N translocation to the seeds. The advantage of using urea is that it facilitates the accumulation of other nutrients such as Mn^{2+} and permits the transport of nutrients through a more permeable cell cuticle (Malavolta, 1981).

The objective of the present study was to determine the amounts and methods of N application (foliar and soil) to common bean (*Phaseolus vulgaris* L.) which could enhance N_2 fixation and yield. Two field experiments were conducted in locations differing in native rhizobia number and soil N, to give an idea of how these two factors might affect bean response to N fertilizer.

Material and methods

The first experiment (Site 1) was initiated in May 1987 at the experimental field of the Escola Superior de Agricultura "Luiz de Queiroz", São Paulo, in a soil classified as Oxic Paleudalf (loamy), pH (H_2O) 5.2; O.M. (%) – 3,98; P (resin method) – 48 ppm; K, Ca, Mg and S (meq. 100 cm^{-3}) – 0.11, 3.29, 1.34 and 5.44, respectively; and V – 60.1%. The area was amended with 2 t ha^{-1} of lime to increase the soil pH up to 5.8, 30 kg P_2O_5 ha^{-1} of simple superphosphate and 600 g ha^{-1} of B. The second trial (Site 2) was initiated in November 1987, at the Experimental Station of EMBRAPA – Fazenda Canchim, São Carlos, São Paulo, in a soil classified as Ultisol (silty loam), pH (H_2O) 5.5; P (resin method) – 2 ppm; K, Ca, Mg, H + Al, S and C.E.C. (meq. 100 cm^{-3}) – 0.12, 1.4, 0.6, 3.4, 2.1 and 5.5, respectively, and V – 38%. The field received 120 kg P_2O_5 ha^{-1} of simple superphosphate, 30 kg K_2O and 1 kg ha^{-1} of a combined formula of micronutrients from a commercial source. Average temperatures at Site 1 ranged from 18°C to 25°C (fall) during the vegetative growth of bean and Site 2 from 23°C to 28°C (spring). The two sites contained different numbers of native *Rhizobium*. Site 1 which is located in an area where bean is cultivated often, had 10^6 cells *Rhizobium* g^{-1} soil. Site 2 had not been planted recently to legumes and therefore had few if any indigenous rhizobia.

At both sites, two set of plots were used for ^{14}N and ^{15}N determinations using ^{15}N labelled or unlabelled fertilizer. The plot sizes were 2 × 2 and 3 × 4 m for the ^{15}N labelled and unlabelled plots, respectively, randomized in three blocks. Seeds were sown in lines with 0.5 m interrow spacing and density of 300,000 plants ha^{-1}. Miniplots of 1.5 × 1.5 m were marked inside the ^{15}N plots for the N treatments. The ^{15}N-labelled treatments are shown in Table 1. Rice was used as a non-fixing reference crop. ^{15}N-labelled urea with 5, 2 and 1 atom% excess was applied to the 10, 30 and 50 kg N ha^{-1} treatments, respectively.

A mixture of strains CM-05 (CENA-USP, Piracicaba, São Paulo) and UMR-1899 (originally provided by Dr. P.H. Graham, University of Minnesota, U.S.A.) grown separately in gamma sterilized commercial peat prepared at CENA was added onto the seeds, previously sprayed with 10% sucrose solution, and pelleted with calcium carbonate at the rate of 35 g inocula and 50 g $CaCO_3$ per kg bean seeds. Untreated seeds were used in the uninoculated treatment (control).

N-urea was added in two forms: (1) soil and (2) sprayed on foliage. The N was applied at three times – a) sowing (0 days after emergence – DAE), b) after the plants had formed the third leaf (21 DAE), and c) early flowering stages (35 DAE) (Table 1). All treatments received an addition of 10 kg N ha^{-1} (labelled or unlabelled) at sowing to standardize N

Table 1. Method and rate of N fertilizer application

	Treatment	Soil at planting	Foliage after emergence
Bean	10*	–	10
Bean	10	–	10*
Bean	10*	–	30
Bean	10	–	30*
Bean	10*	–	50
Bean	10	–	50*
Bean	10*	10	–
Bean	10	10*	–
Bean	10*	30	–
Bean	10	30*	–
Bean	10*	50	–
Bean	10	50*	–
Rice	40*	–	–

* Added as urea and labelled with 5%, 2% and 1% atom excess in the 10, 30 and 50 kg N ha^{-1} bean treatments. Urea for rice was labelled with 1% atoms in excess.

treatments. Urea was added to the soil or sprayed early in the morning (to avoid burning the leaves and minimize losses due to evaporation) to the foliage at a rate of 10 kg N ha^{-1} diluted to a concentration of 3% urea for each time.

At the R3 growth stage (42 days after emergence – DAE, 50% bloom), 10 plants from each treatment in unlabelled plots were harvested to obtain data on ARA (Site 1), weight (Sites 1 and 2) and nodule number (Site 2). A complete determination of shoot dry matter, and N and P content was performed in plants harvested at the R3 stage in the unlabelled plots of Site 2. The final harvests of the ^{15}N-labelled plots were made at 92 and 96 DAE when seed yield, seed nitrogen and phosphorus (particularly for Site 2) were measured. ^{15}N determination were made by mass spectrometer Atlas-Varian model CH-4 following Dumas combustion of plant materials (Proksch, 1969).

Results

There were few native rhizobia at Site 2 as shown by uninoculated plants which had an average of 1.8 nodules and 2.6 mg per plant nodule weight (Table 2). A highly significant increase in nodule weight (63.7 mg nodule per plant) was obtained with inoculation at this site. Native rhizobia were active at Site 1, producing an average of 65 mg of nodules per plant. Also, Site 1 contained more available P than Site 2, allowing better survival of naturally established rhizobia (Saito and Ruschel, 1978).

Nitrogen added at low rates (10 kg N ha^{-1} at sowing) stimulated nodule weight (Sites 1 and 2) and to a less extent nodule number (Site 2). The favorable effect of small amounts of N was seen also for ARA at Site 1, where an increase of 2.35 μM C_2H_4 $plant^{-1}$ h^{-1} to 9.12 μM C_2H_4 $plant^{-1}$ h^{-1} was observed with additions of 10 kg N ha^{-1} at sowing. Starter nitrogen had less effect on nodulation than rhizobial inoculation at Site 2. When N was added after emergence, both method and rate affected nodulation patterns. High rates, up to 40 kg N ha^{-1} reduced nodule weight, although not as much as nodule number (Site 2). N added directly to the soil affected nodule mass and activity (ARA, Site 1), especially when higher rates were added at early stages of plant growth (30 kg N ha^{-1}).

Rhizobium inoculation alone increased plant growth significantly ($p < 0.05$) (Table 3), with increases up to 57% primarily in shoot biomass

Table 2. Effect of N application methods on nodulation and dry matter production (shoot and root) of common bean. Sites 1 and 2

Treatment	Nodulation Site 1		Site 2	
	(no/plant)	(mg/plant)	(no/plant)	(mg/plant)
Control	38.0 a*	65.0 bc	1.8 b	2.6 d
Inoculated	36.3 a	74.5 abc	37.3 a	63.7 ab
10 P	62.5 a	124.8 a	39.4 a	66.1 a
10 P + 10 F	64.3 a	99.9 ab	44.6 a	64.1 ab
10 P + 30 F	33.2 a	70.0 bc	44.0 a	37.3 c
10 P + 50 F	50.6 a	62.0 bc	37.7 a	40.2 bc
10 P + 10 S	50.3 a	92.5 abc	32.7 a	51.6 abc
10 P + 30 S	47.1 a	47.7 bc	32.5 a	37.4 c
10 P + 50 S	33.9 a	142.3 c	37.3 a	52.3 abc
10 P + 50 S + 20 F	38.1 a	44.1 c	31.7 a	34.4 c
30 S	40.5 a	71.7 abc	28.9 a	43.1 abc
CV (%)	37	39	44	34

* Means followed by same letters are not significant. Tukey test ($p < 0.05$).
10 P = 10 kg N ha^{-1} applied to the soil at sowing.
10 S, 30 S and 50 S = kg N ha^{-1} applied to the soil.
10 F, 30 F and 50 F = kg N ha^{-1} applied on the leaves.

Table 3. Effect of N fertilization methods on dry matter, production (shoot and root) of common bean plants, at mid-flowering stage. Site 2

Treatment	Shoot (S)	Root (R)	Total (S + R)	Relative increase (%)
	($kg\ ha^{-1}$)			
Control	1263 d*	123.6 ab	1386 d	–
Inoculated	2037 ab	141.3 ab	2178 ab	57
10 P	1591 cd	122.3 b	1712 cd	23
10 P + 10 F	2215 cd	130.0 ab	2345 a	69
10 P + 30 F	1762 bc	137.1 ab	1899 bc	37
10 P + 50 F	2126 ab	143.3 a	2269 ab	63
10 P + 10 S	2089 ab	143.7 a	2232 ab	61
10 P + 30 S	2049 ab	140.7 a	2189 a	58
10 P + 50 F	2001 ab	134.7 ab	2136 ab	54
10 P + 30 S + 20 F	1820 abc	126.5 ab	1947 abc	40
30 S	1946 abc	127.4 ab	2076 abc	50
C.V. (%)	9	13	13	

* Means followed by same letters are not significant. Tukey test ($p < 0.05$).
10 P = 10 $kg\ N\ ha^{-1}$ applied to the soil at sowing.
10 S, 30 S and 50 S = $kg\ N\ ha^{-1}$ applied to the soil.
10 F, 30 F and 50 F = $kg\ N\ ha^{-1}$ applied on the leaves.

when compared to the control. At this stage, 10 $kg\ N\ ha^{-1}$ added at sowing was not enough to stimulate plant growth. In addition, N_2 fixation promoted by inoculated rhizobia contributed significantly more N compared to treatments with higher amounts of N. The highest biomass value was obtained in the treatment with 20 $kg\ N\ ha^{-1}$, half applied at sowing and half sprayed on foliage.

Yield, and the N and P accumulated in seeds were primarily affected by seed inoculation (Table 4), although higher values of these parameters were obtained at the highest N rate. Additions of 10 $kg\ N\ ha^{-1}$ at sowing gave values comparable to other higher N treatments.

Yield (Fig. 1A) and nitrogen accumulated in seeds (Fig. 1B) from the labelled plots of both sites are compared. Only the rates applied after sowing (10, 30 and 50 $kg\ N\ ha^{-1}$) are shown. Yield and seed-N from Site 2 were higher than Site 1 on the average, where significant differences among treatments were also found. Increasing rates of N when added directly to soil had little effect on seed production.

Foliage spray significantly increased seed N independently of N treatments. When the 30 $kg\ N\ ha^{-1}$ rate was applied as foliar application a suppressive effect (significant in Site 2) was observed on the seed yield and consequently seed-N. This could be attributed to a deleterious effect of N on both enzyme activities – nitrogen-

Table 4. Effect of N fertilization methods on N and P contents in common bean plants, at mid-flowering stage. Site 2

Treatment	Total N	Total P
	($kg\ ha^{-1}$)	
Control	31.4 d*	4.7 c
Inoculated	70.1 ab	8.9 a
10 P	52.5 c	6.4 bc
10 P + 10 F	74.5 ab	8.9 a
10 P + 30 F	62.2 bc	7.2 ab
10 P + 50 F	77.1 ab	8.9 a
10 P + 10 S	69.5 ab	8.9 a
10 P + 30 S	73.5 ab	8.2 ab
10 P + 50 F	73.6 ab	8.7 ab
10 P + 30 S + 20 F	69.2 ab	7.6 ab
30 S	74.9 ab	8.9 a
C.V. (%)	13	18

* Means followed by same letters are not significant. Tukey test ($p < 0.05$).
10 P = 10 $kg\ N\ ha^{-1}$ applied to the soil at sowing.
10 S, 30 S and 50 S = $kg\ N\ ha^{-1}$ applied to the soil.
10 F, 30 F and 50 F = $kg\ N\ ha^{-1}$ applied onto the leaves.

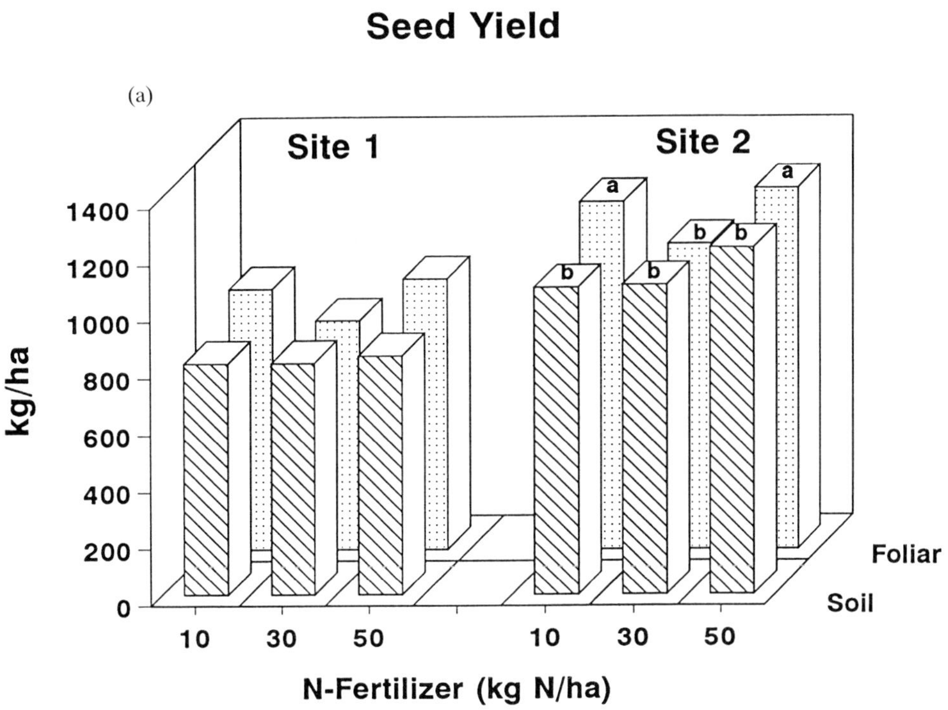

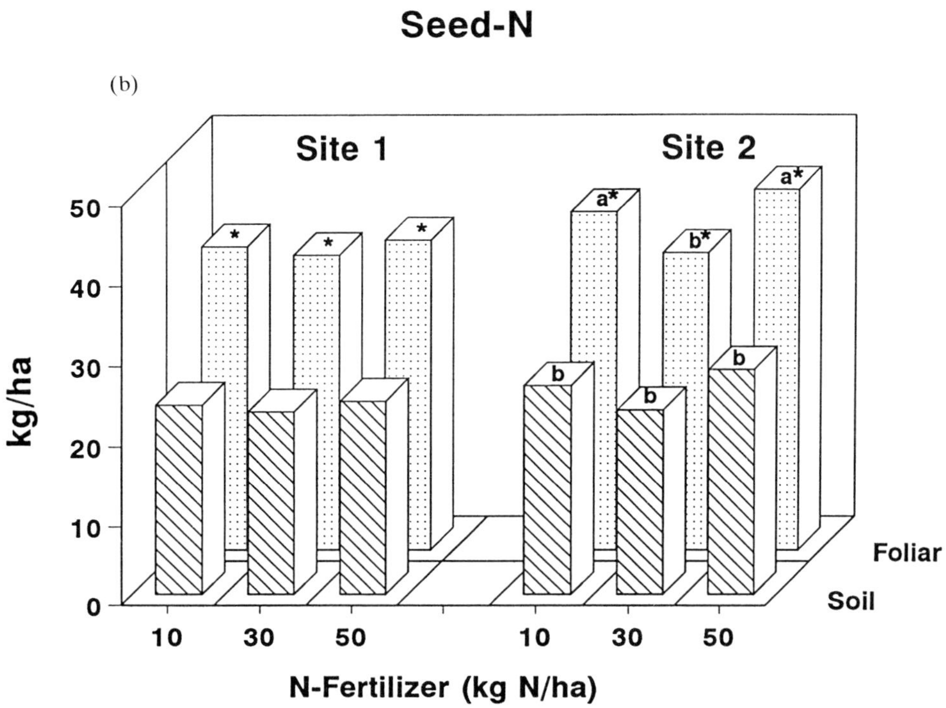

Fig. 1. Yield (**A**) and N content (**B**) of bean seeds in response to foliar (F) or soil N-applications (S) at two experimental sites.

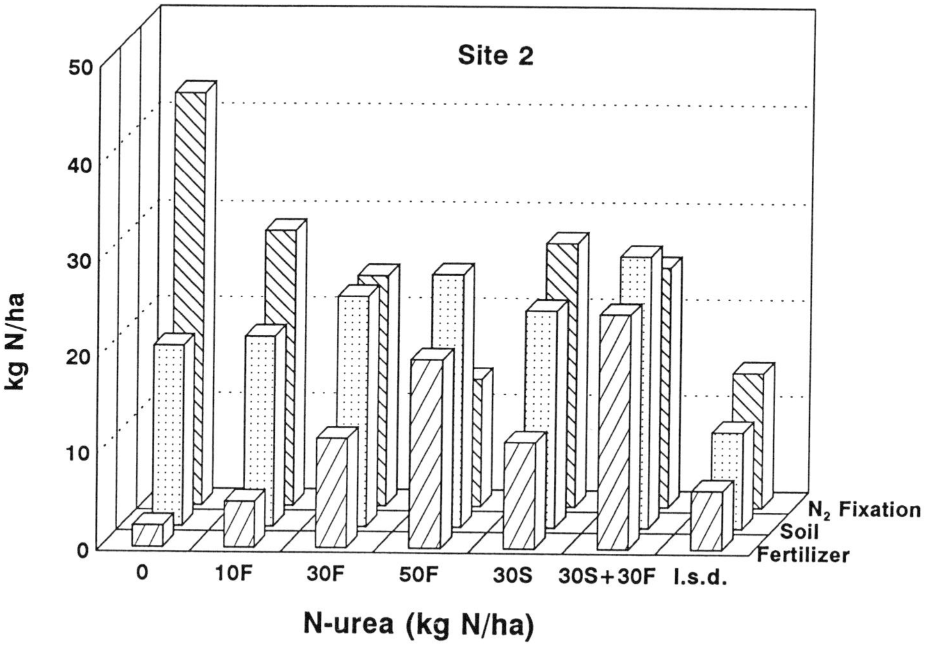

Fig. 2. Amounts of nitrogen derived from N_2 fixation, soil + fertilizer estimated in bean seeds, in response to increasing levels (10, 30 and 50 kg N ha^{-1}) and two application methods of N-urea – foliar (F) and soil (S). (Site 2)

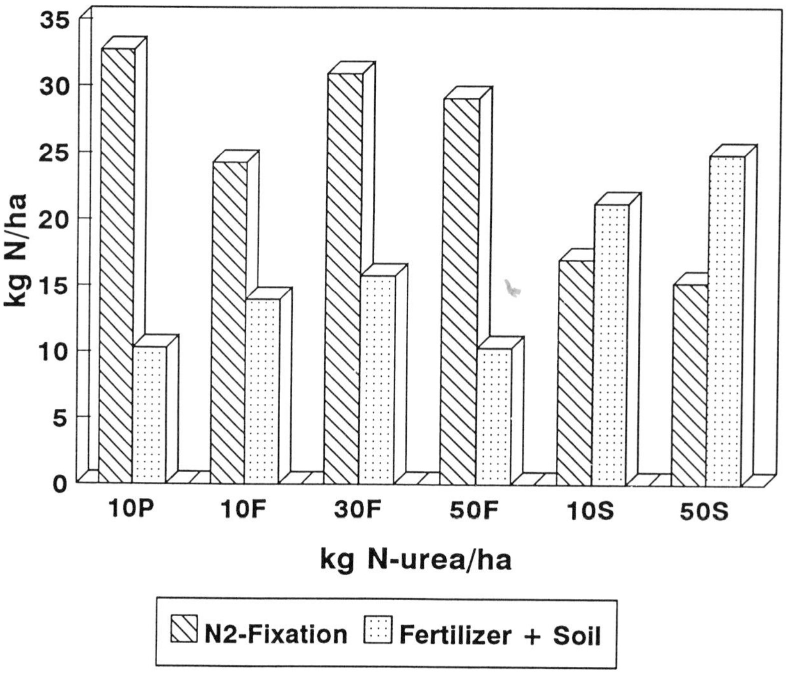

Fig. 3. Amount of nitrogen from N_2 fixation, fertilizer and soil in bean seeds in response to increasing levels of N-urea (10, 30 and 50 kg N ha^{-1}) and two application methods – foliar (F) and soil (S).

ase (more active at low N rates) and nitrate reductase (more active at high N rates). Both enzymes may be partially inhibited at this intermediate N rate, affecting to some extent the yield and N accumulation in the seed.

The proportion of fixed N_2 was correlated to nodulation, as decreased values were observed with increasing rates of N (Fig. 2). In the presence of 10 kg N ha^{-1}, N_2 fixation was highest in Site 1 (82.1%) and Site 2 (67.8%). Lowest N_2 fixation rates were found in treatments with high N. The proportion of N in seeds from fertilizer gave maximum values of 25.4% and 28.2% for the highest soil and foliar N application rates.

Estimation of amount of N_2 fixed (Fig. 3) gave values ranging from 14.9 to 31.7 kg N ha^{-1} (Site 1) and 13.2 to 42.7 kg N ha^{-1} (Site 2). The three-fold increase in the amounts of fixed N_2 in Site 2 indicates that it is possible to obtain significant amounts of nitrogen from N_2 fixation by using selected *Rhizobium* in sites where native rhizobia are not predominant, confirming previous data collected from sites containing different native rhizobia number (Saito, 1982).

Discussion

In the present study, yield was influenced more by foliar than soil N applications. Most significantly, all N foliar treatments contributed to increased seed N in both sites. Our data on seed-N and uptake of fertilizer N indicate that N translocation is more efficient when bean plants received N-urea directly from the leaves.

Previous reports from other grain legumes indicate that the leaves are the principal tissue for the uptake and assimilation of exogenous N (Ohyama, 1973; Pate, 1980). Our data from Site 2 (Fig. 2) indicate that N added to soil (30 kg N ha^{-1}) had little effect on the total N accumulated in the plant when compared to the same N rate added through foliage. Similar results were observed previously in soybean and cowpea, which showed yield responses to supplemental N added at pod-filling when N demand is greater (Franco et al., 1979) and root uptake of N is more active than at the initial growth stages (Eaglesham et al., 1983; Lawn and Brun, 1983; Afza et al., 1987).

DuBois and Burris (1986) suggested that in common beans, N_2 fixation at and beyond the R3 stage furnishes N directly to reproductive parts (flower/fruits). Uptake of other forms of N results in large percentages of assimilated N from vegetative parts (leaflets) during pod fill. Recent data showed a direct accumulation of fixed N_2 to bean flowers/fruits, translocated as ureides from below ground organs (Thomas et al., 1988).

In addition to the facilitated C-transport to seeds through ureides, the deleterious effect of N fertilizer to nodules on secondary roots could be alleviated by avoiding N additions directly to the soil, since there are significant contributions of lateral root nodules to the nitrogen pool in the common bean (Wolyn et al., 1989) and soybean (Hardarson et al., 1989; Wadisirisuk et al., 1989). From our results, we can conclude that as found in other crops, bean is capable of fixing atmospheric N in the presence of fertilizer N applied in small amounts to the leaves, resulting in increased yield and seed N.

Acknowledgements

This contribution was supported by the U.S. National Academy of Sciences – BOSTID (BNF-BR-4-86-53) and IAEA (311-D1-BRA-4249) research grants. The authors would like to thank Dr. Takashi Muraoka for scientific support. S.M.T. acknowledges funding from Conselho Nacional de Pesquisa e Desenvolvimento (Brazil).

References

Afza R, Hardarson G and Zapata F 1987 Effects of delayed soil and foliar N fertilization on yield and N_2 fixation of soybean. Plant and Soil 97, 361–368.

DuBois J D and Burris R H 1986 Comparative study of N uptake and distribution in three lines of common bean (*Phaseolus vulgaris* L.) at early pod filling stage. Plant and Soil 93, 79–86.

Dunphy E J and Hanway J J 1976 Water-soluble carbohydrate accumulation in soybean plants. Agron. J. 68, 697–700.

Eaglesham A R J, Hassouna S and Seegers R 1983 Fertilizer-N effects on N_2 fixation by cowpea and soybean. Agron. J. 75, 61–66.

Franco A A, Pereira J C and Neyra C A 1979 Seasonal

patterns of nitrate reductase and nitrogenase activities in *Phaseolus vulgaris* L. Plant Physiol. 63, 421–424.

Hardarson G, Zapata F and Danso S K A 1984 Effect of plant genotype and nitrogen fertilizer on symbiotic nitrogen fixation by soybean cultivars. Plant and Soil 82, 397–405.

Harper J E 1974 Soil and symbiotic nitrogen requirements for optimum soybean production. Crop Sci. 14, 255–260.

Frith G J T and Dalling M J 1980 The role of peptide hydrolases in leaf senescence. *In* Senescence in Plants. Ed. K V Thimann. pp 117–130. CRC Press Inc., Boca Raton, FL.

Hardarson G, Golbs M and Danso S K A 1989 Nitrogen fixation in soybean (*Glycine max* L. Merrill) as affected by nodulation patterns. Soil Biol. Biochem. 21, 783–787.

Lawn R J and Brun W A 1974 Symbiotic nitrogen fixation in soybeans. III. Effect of supplemental nitrogen and intervarietal grafting. Crop Sci. 14, 22–25.

Machado J R, Rosolem C A, Balducci Jr J J and Nakagawa J 1982 Foliar fertilization in common bean. I. Estudo de epocas de aplicacao de nitrogenio. Turrialba 32, 417–421.

Malavolta E 1981 Elementos de Nutricao de Plantas. Ed. Ceres, pp 80–94, São Paulo.

Miysaka S, Freire E S and Mascarenhas H S 1963 Modo e epoca de aplicacao de nitrogenio na cultura do feijoeiro. Bragantia 22, 511–519.

Muraoka T and Neptune A M L 1978 Efeito da aplicacao foliar de polifosfato, superfosfato, ureia e yogen na producao de feijoeiro (*Phaseolus vulgaris* L.). Rev. Bras. Ci. Solo 2, 51–55.

Ohyama T 1983 Comparative studies on the distribution of nitrogen in soybean plants supplied with N_2 and NO^{-3} at the pod filling stage. Soil Sci. Plant Nutr. 29, 133–145.

Pate J S 1980 Transport and partitioning of nitrogenous solutes. Annu. Rev. Plant Physiol. 31, 313–340.

Proksch G 1969 Routine analysis of ^{15}N in plant material by mass-spectrometry. Plant and Soil 31, 380–384.

Ramon Garcia and Hanway J J 1976 Foliar fertilization of soybeans during the seed-filling period. Agron. J. 68, 653–657.

Rosolem C A, Machado J R, Balducci Jr J J and Liem T H 1982 Adubacao foliar do feijoeiro. III. Efeitos do nitrogenio com e sem cobertura nitrogenada. Turrialba 32, 423–428.

Saito S M T and Ruschel A P 1978 Influence of lime, phosphorus fertilizer and micronutrients in the natural nodulation of beans (*Phaseolus vulgaris* L.). An. Esc. Sup. Agric. Luiz de Queiroz 25, 545–556.

Thomas R J and Hungria M 1988 Effect of K on N_2 fixation nitrogen transport and N harvest index of beans (*Phaseolus vulgaris* L.). J. Plant Nutr. 11, 175–188.

Vasilas B L, Legg J O and Wolf D C 1980 Foliar fertilization of soybeans. Absorption and translocation of ^{15}N-labelled urea. Agron. J. 72, 271-275.

Wadisirisuk P, Danso S K A, Hardarson G and Bowen G D 1989 Influence of *Bradyrhizobium japonicum* location land movement on nodulation and nitrogen fixation in soybeans. Appl. Environ. Microbiol. 55, 1711–1716.

Wolyn D J, Attewell J, Ludden P W and Bliss F A 1989 Indirect measures of N_2 fixation in common bean (*Phaseolus vulgaris* L.) under field conditions: The role of laterial root nodule. Plant and Soil 113, 181–187.

Plant and Soil **152**: 131–138, 1993.

PLSO BNF14

Minimizing the effect of mineral nitrogen on biological nitrogen fixation in common bean by increasing nutrient levels

S.M. TSAI[1], R. BONETTI[1], S.M. AGBALA[2] and R. ROSSETTO[1]
[1]*Centro de Energia Nuclear na Agricultura, University of São Paulo, Caixa P.O. Box 96, Piracicaba CEP 13400-970, São Paulo, Brazil*; [2]*Commissariat Général à L'Energie Atomique, Kinsasha, Zaire*

Key words: bean, ^{15}N isotope, nitrogen fixation, N fertilizer, phosphorus, potassium, *Rhizobium tropici*, sulphur

Abstract

Although common bean (*Phaseolus vulgaris* L.) has good potential for N_2 fixation, some additional N provided through fertilizer usually is required for a maximum yield. In this study the suppressive effect of N on nodulation and N_2 fixation was evaluated in an unfertile soil under greenhouse conditions with different levels of soil fertility (low = no P, K and S additions; medium = 50, 63 and 10 mg kg^{-1} soil and high = 200, 256 and 40 mg kg^{-1} soil, respectively) and combined with 5, 15, 60 and 120 mg N kg^{-1} soil of ^{15}N-labelled urea. The overall average nodule number and weight increased under high fertility levels. At low N applications, nitrogen had a synergistic effect on N_2 fixation, by stimulating nodule formation, nitrogenase activity and plant growth. At high fertility and at the highest N rate (120 mg kg^{-1} soil), the stimulatory effect of N fertilizer on N_2 fixation was still observed, increasing the amounts of N_2 fixed from 88 up to 375 mg N plant^{-1}. These results indicate that a suitable balance of soil nutrients is essential to obtain high N_2 fixation rates and yield in common beans.

Introduction

Common bean is capable of fixing large amounts of atmospheric N_2 under certain conditions, but usually some mineral N is needed to achieve substantial yields under the prevailing cropping system in Latin America. However, nitrogen fertilization affects nodulation of bean plants and therefore the usually-recommended rates of 40–60 kg N ha^{-1} suppress N_2 fixation (Ruschel et al., 1979a; Graham, 1981). On the other hand, low rates have been shown to enhance nodule formation and function but are not sufficient to achieve maximum yields (Sistachs, 1970; Rosas and Bliss, 1987). While it is known that bean plants require high levels of P, K, S, Ca, Mg, Mo, Co and B (Bonetti et al., 1984; Franco and Day, 1980; Ruschel and Reuszer, 1973; Saito and Ruschel, 1978), production is restricted mostly to marginal soils where sufficient nutrients are not available. In soybean and tropical legumes, nodulation is especially sensitive to non-optimal levels of combined nutrients (Cassman et al., 1981; França et al., 1973; Gates and Wilson, 1974; Hernandez and Focht, 1985; Lynd and Ansman, 1989, 1990).

Gates and Muller (1979) observed that nodulation in soybean is affected by unbalanced nutritional conditions of N, P and S, and that N_2 fixation was not affected by mineral N only in presence of high levels of P and S. Lynd and Ansman (1989) found decreases in nodule number of peanut when K was applied alone, but not when P and Ca were added. A stimulatory effect of the high rate of 60 mg kg^{-1} of NH_4NO_3 on nodulation also has been reported in presence of P (Gates and Wilson, 1974).

This study was performed to determine whether applied nitrogen affects nodulation and plant biomass under limited and balanced soil nutrient situations. In addition, the amounts of N_2 fixed with increasing soil fertility were compared to determine if a nutritional balance for the plant growth could alleviate the detrimental effect of N fertilizer on the biological process.

Material and methods

Bean plants (*Phaseolus vulgaris* L.) were grown in a greenhouse in pots of 3 kg of an acidic dark oxissol collected from a 'cerrado' (savanah) region of Sao Paulo State (22°42′30″ S latitude and 47°38′00″ W longitude), Brazil. Seeds of cv. Carioca were provided by Dr. Eduardo Bulisani, Instituto Agronomico de Campinas – S. Paulo. Prior to sowing, seeds were surface sterilized with 0.5% HOCl for 5 min, rinsed several times with distilled water and inoculated with 0.10 mL (approximately 10^7 cells seed^{-1}) of a 3-day-old yeast mannitol broth culture of the *Rhizobium tropici* strain UMR-1899 (originally from Dr. P.H. Graham, University of Minnesota-MN, U.S.A.), known to be effective with common bean in tropical conditions (Vargas and Graham, 1989, Martinez-Romero *et al.*, 1991). Plants were thinned to two per pot at about 7 days after emergence (DAE).

Experimental pots were fertilized according to the usual agricultural recommendations to give a low, medium and a high level of phosphorus, potassium and sulphur (Table 1). ^{15}N-labeled urea (4.763 atom % ^{15}N) was added at planting at the rates of 15 and 60 mg N kg^{-1} soil, according to the low, medium and high N-rate treatments. An additional supply of 45 and 60 mg N kg^{-1} soil was added 3 weeks after germination for the treatments of 60 and 120 mg N kg^{-1} soil.

Chemical analysis of the soil gave the following values: pH ($CaCl_2$) – 5.5; O.M. – 3%; P – 10 μg g^{-1} soil (resin); K – 0.18, Ca – 2.11, Mg – 0.94, H + Al – 1.34 meq 100 cm^{-3}, respectively. The soil was air dried and sieved (2 mm) and pH corrected to 6.0 with addition of 1.0 g of calcium carbonate per pot.

A randomized complete block design with four replicates was used at each sampling time (42 DAE and 90 DAE). The first sampling determined nodulation parameters – number, mass and acetylene reduction activity (ARA) – of the plants at the R_3 growth stage (42 DAE). For acetylene reduction activity (ARA) determinations gas samples were injected into a GC-65 Beckman gas chromatograph, with an H_2 flame ionization detector and N_2 as flow gas. At the second sampling at 90 DAE plant biomass (shoot plus pods) and total N, using a Perkin-Elmer autoanalyser were evaluated. ^{15}N enrichment of plant materials was determined for the two samplings, after Dumas combustion (Proksch, 1969) in an Atlas-Varian model CH-4 mass spectrometer. ^{15}N calculations were done as shown by Danso et al. (1983).

Table 1. Amounts of P, K and S added to the soil to obtain low, medium and high fertility levels and N-fertilizer rates used in this study

Fertility level	N*	P	K	S
	(mg kg^{-1} soil)			
Low	5	–	–	–
Low	15	–	–	–
Low	60	–	–	–
Low	120	–	–	–
Medium	5	50	63	10
Medium	15	50	63	10
Medium	60	50	63	10
Medium	120	50	63	10
High	5	200	256	40
High	15	200	256	40
High	60	200	256	40
High	120	200	256	40

* The application of 60 and 120 mg N kg^{-1} soil was split, at sowing and 3 weeks after germination, corresponding to 15 + 45 and 60 + 60 mg N kg^{-1} soil, respectively.

Results

Nodulation and plant growth

At 42 DAE, significant increases of nodule number, mass and acetylene reduction values were obtained by increasing soil fertility (Table 2). Nodule numbers increased significantly with increasing fertility from 66 (low fertility) to 168 (medium), and to 222 (high) nodules plant^{-1}.

Table 2. Nodule number, mass and acetylene reduction activity (ARA) of bean plants (42 DAE) grown under increasing soil fertility

	Fertility level		
	Low	Medium	High
Number (No. of plant^{-1})	66 C	168 B	222 A
Mass (mg plant^{-1})	39 B	111 A	108 A
ARA (μMol C_2H_4 pl^{-1} h^{-1})	6.4 B	14.5 A	17.5 A

* Capital letters may be used for comparisons among soil treatments.
Means followed the same letters do not differ statistically by Duncan's Multiple Range Test ($p < 0.05$).
Low = no P, K and S additions.
Medium = 50, 63 and 10 mg kg^{-1} soil of P, K and S
High = 200, 256 and 40 mg kg^{-1} soil of P, K and S

Although nodule mass nearly doubled from low – (39 mg plant^{-1}) to medium fertility (111 mg plant^{-1}), this trait as well as ARA was influenced less by additions of higher amounts of minerals. Nitrogenase activity was also increased by the P, K and S additions at 42 DAE, indicating that nodule function also depends on the nutritional status of the soil in order to obtain a substantial amount of fixed N_2.

Nitrogen showed a dual effect on nodulation (Fig. 1A, B, C and D). At low fertility, the addition of 60 mg N kg^{-1} soil was detrimental to nearly all nodulation traits (nodule number, mass and ARA, and to a less extent nodule size). However, this effect was less as the concentration of P, K and S in the soil was increased. Nodule number increased significantly from 41 (low) to 151 (medium) and 310 (high) nodules per plant. All nodulation parameters were affected greatly by the highest rate of N fertilizer (120 mg kg^{-1} soil). The synergistic effect of 15 mg N kg^{-1} soil observed at low fertility was also observed with higher N rates (60 mg N kg^{-1} soil) under higher soil fertility, giving the highest values for all nodulation parameters.

The high ARA rates obtained under the high fertility condition were comparable to ARA of soybean grown under similar conditions (Ruschel et al., 1979b). This was observed also in soybean by Gates and Muller (1979) where no detrimen-

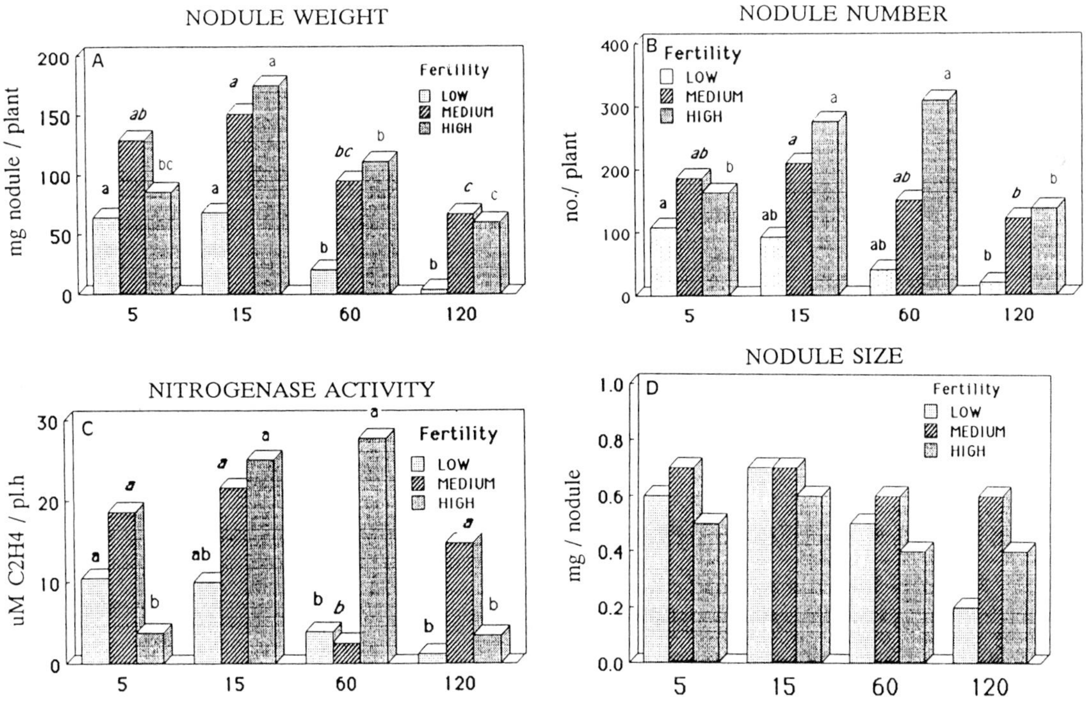

Fig. 1. Nodulation traits evaluated in N treatments (5, 15, 60 and 120 mg N kg^{-1}) of bean plants influenced by P, K and S additions: a) nodule weight (mg nodule/plant), b) nodule number (per plant), c) nitrogenase activity of nodules (μMol C_2H_4 plant^{-1} h^{-1}) and d) average nodule size (mg nodule^{-1}). Comparisons are made only among N treatments from each soil.

tal effect of N fertilizer as high as 120 mg kg^{-1} was observed when high rates of P and S were used. Although not statistically different, high averages in nodule size (Fig. 1D) were obtained at low N and at high fertility. It appears that bean can still develop nodules at relatively high N concentrations when grown in fertile soil. Gates and Wilson (1974) have shown that 60 ppm N were less suppressive to nodulation in *Stylosanthes humilis* in the presence of medium and high rates of P fertilizer.

At 90 DAE, increases in dry matter and N content due to P, K and S additions were more consistent than those due to N application (Table 3). As P, K and S increased, dry matter increased from 6.1 g $plant^{-1}$ to 7.7 g $plant^{-1}$ (medium fertility) and to 10.7 g $plant^{-1}$ (high fertility). Total N followed the same pattern as dry matter, doubling the values with P, K and S additions (258.8 mg N $plant^{-1}$ in contrast to 127.3 mg N $plant^{-1}$ of the control treatment).

N_2-fixation and fertilizer use

The proportion of N derived from air (%Ndfa) was highly correlated to nodulation and total N (Fig. 2). The average proportion of N_2 fixed was 51.9% at low fertility and increased significantly to 66% and 64.7% ($p < 0.05$) at medium or high fertility, respectively. Although not statistically significant, overall highest N_2 fixation rates were observed with low inputs of N fertilizer (up to 15 mg N kg^{-1} soil) at medium soil fertility conditions.

The uptake of N fertilizer by common bean was moderate at low fertility levels (Fig. 2), and was increased only with additions of 120 mg N kg^{-1} soil at higher soil fertility levels. The low response to N fertilizer indicates that soil factors other than nitrogen could limit bean plant growth and productivity in marginal soils. Soil nitrogen was also more available to the plant when P, K and S were present in high amounts.

In general, N fertilizer use averaged 35–50%, reaching the highest values of around 65% at an intermediate fertility (Table 4). The depressive effect of high levels of N on nodulation was alleviated at high fertility levels, as shown by the N_2 fixed values (Table 4). An estimation of the total N_2 fixed gave values corresponding to 312 and 375 mg N $plant^{-1}$, in treatments when 60 and 120 mg N kg^{-1} soil were added, in contrast to 128 and 88 mg N $plant^{-1}$ under low, and 187

Table 3. Dry matter and N content of bean plants at 42 DAE and 90 DAE grown under increasing levels of P, K and S and N-urea

N rate*	42 DAE			90 DAE		
	Low	Medium	High	Low	Medium	High
	Dry matter (g $plant^{-1}$)					
5	1.9	3.5	3.5	6.2	8.1	11.6
15	1.9	3.9	4.2	5.6	8.1	8.7
60	2.8	3.9	4.5	5.9	7.0	10.4
120	2.3	5.1	4.9	6.7	7.4	12.3
Mean	2.2 B	4.1 A	4.3 A	6.1 C	7.7 B	10.7 A
	Total nitrogen (mg N $plant^{-1}$)					
5	46.4	96.8	92.6	92.9	145.1	264.2
15	47.4	107.9	109.4	143.0	193.4	173.7
60	72.9	101.5	112.0	145.5	139.7	268.0
120	79.0	128.7	116.7	127.7	195.9	328.1
Mean	61.4 B	108.7 A	107.7 A	127.3 C	164.4 B	258.8 A

* mg N kg^{-1} soil
Low = no P, K and S additions.
Medium = 50, 63 and 10 mg kg^{-1} soil, respectively.
High = 200, 256 and 40 mg kg^{-1} soil, respectively.
Means followed by the same letters within each sampling time do not differ significantly by Duncan's Multiple Range Test ($p < 0.05$).

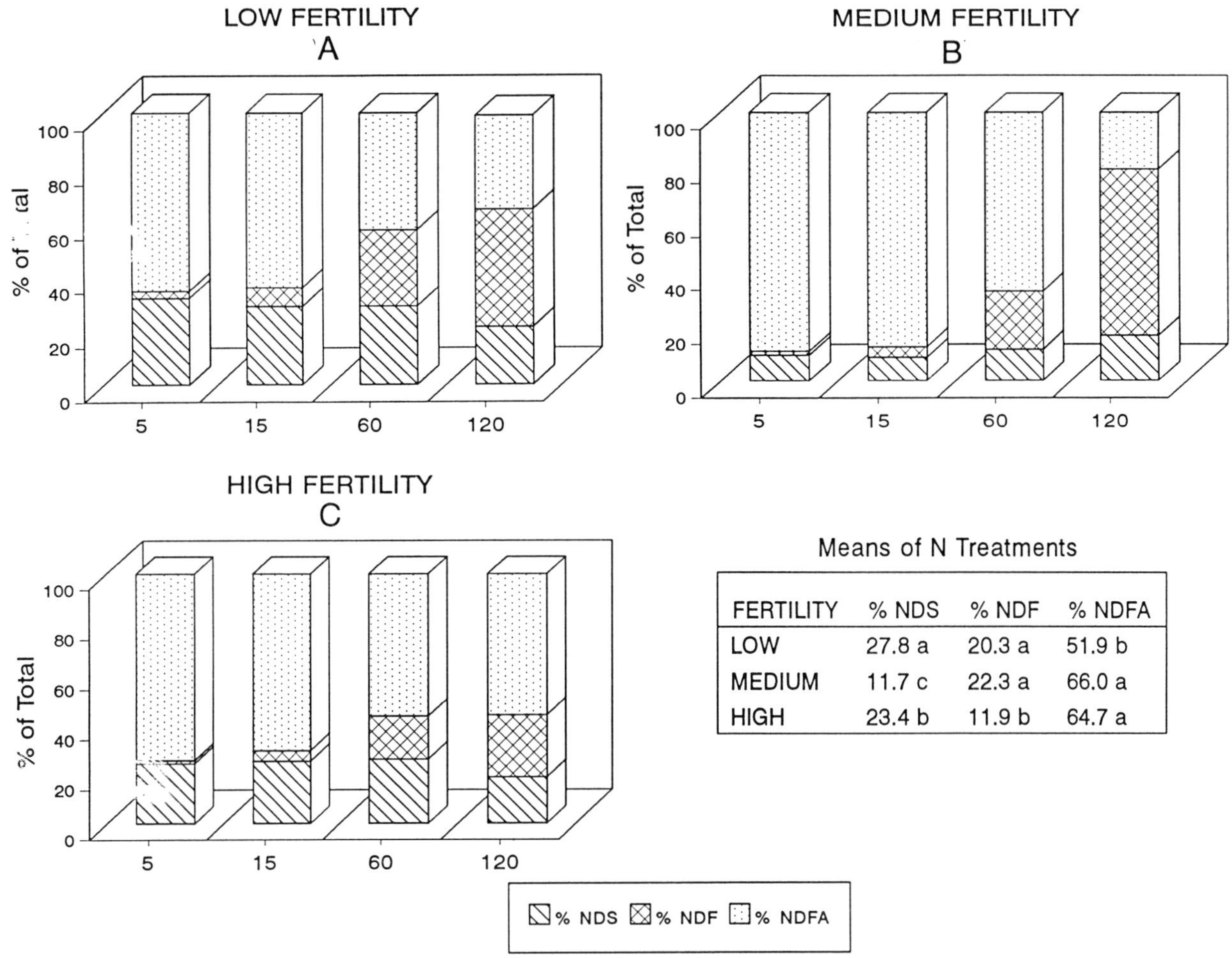

Means of N Treatments

FERTILITY	% NDS	% NDF	% NDFA
LOW	27.8 a	20.3 a	51.9 b
MEDIUM	11.7 c	22.3 a	66.0 a
HIGH	23.4 b	11.9 b	64.7 a

Fig. 2. Proportion (% of total) of N from soil, fertilizer and N_2 fixation in bean plants at 90 DAE, as function of N fertilizer additions (5, 15, 60 and 120 mg N kg^{-1}) and grown under three soil fertility levels. Mean data of N treatments for the soils are shown on the table inserted in the Figure.

and 87 mg N $plant^{-1}$ under medium fertility conditions (Table 5).

Discussion

Historically, selection in bean breeding programmes in Latin America has been for high yielding cultivars, that are responsive to N fertilizer. Nitrogen application presents a problem for N_2 fixation in common bean, since the recommended amounts of N fertilizer are capable of suppressing nodulation and nodule functioning in this crop (Graham, 1981; Ruschel and Saito, 1977; Silva et al., 1992). However, low amounts of mineral N can stimulate bean nodulation under certain soil conditions (Ruschel et al., 1979a; Vernichenko, 1984). Therefore, the ability of bean to fix enough N_2 for plant growth is a problem of great interest under Brazilian conditions.

Our data show that N_2 fixation can be expressed in a commercial bean cultivar, if inputs of N fertilizer required to increase plant growth are minimal and P, K and S are adequately supplied. By adding P, K and S and a low amount of N fertilizer (urea) substantial increases in nodulation and consequently high amounts of N_2 fixed were obtained with corresponding increases in the plant biomass. Several

Table 4. Effect of different levels of N, P, K, and S on percentage and amount of nitrogen derived from fertilizer (% NDFF and ANDFF, respectively), and efficiency of fertilizer utilization (EFU) by bean plants 90 days after germination

Fertility level*	% NDFF	ANDFF	% EFU
Low	20,3 A	54,1 B	38,0 A
5	2,6 c	4,1 c	32,0 a
15	7,0 c	20,0 c	44,1 a
60	28,3 b	80,9 b	44,9 a
120	43,4 a	111,0 a	30,8 a
Medium	22,3 A	80, A	41,0 A
5	1,5 c	4,4 c	29,2 b
15	4,0 b	15,7 c	34,8 b
60	21,6 b	60,0 b	33,4 b
120	62,2 a	240,1 a	66,7 a
High	11,9 B	67,9 AB	43,1 A
5	1,3 c	7,0 c	46,7 ab
15	4,2 c	14,3 c	31,8 b
60	17,1 b	90,1 b	50,0 a
120	24,8 a	160,2 a	44,5 ab
CV%	20	30	24

Means followed by the same letters do not differ statistically by Duncan Test ($p < 0.05$).
* Low = No P, K, and S addition.
Medium = 50, 63, and 10 mg kg^{-1} of P, K, and S, respectively.
High = 200, 256, and 40 mg kg^{-1} of P, K, and S, respectively.
N: 5, 15, 60 and 120 mg N kg^{-1}.

authors have noted the sensitivity of N_2 fixation to factors such as N fertilizer but others have cited that P and/or S may counteract this detrimental effect (Gates and Wilson, 1974; Gates and Muller, 1979). Under low P levels, nitrogen also can limit the expression of this trait in bean (Pereira and Bliss, 1987, 1989), and there are a few examples of low P tolerance among bean cultivars (Graham and Rosas, 1979). From the few reports of the effect of potassium on N_2 fixation, Thomas and Hungria (1988) have found that this element may contribute to the transport of fixed N_2 from nodules to reproductive organs of beans. P and K together may delay senescence of the soybean nodule, by increasing nodule number and nitrogenase activity (Kungl, 1987). These authors also found synergistic effects of mineral N on nodulation at rates up to 40 mg N kg^{-1} soil. Sulphur also is required to some extent in acidic soils, but can be supplemented partially as a companion ion from commercial P sources.

The high ARA values in this study obtained under a balanced nutritional situation indicate the potential of beans to express N_2 fixing ability in marginal soils, once adequate nutrients are available. Starter nitrogen should be included also in order to assure a high yield. The overall low values obtained for the N fertilizer use (%FUE) when no P, K and S were added, indicate that an additional factor may be limiting N uptake in bean plants under sub-optimum soil conditions. In contrast, as the nutrient level increased all three nitrogen sources became more available to the plant (Fig. 2). These findings indicate that when there are limitations such as P, K or S on the supply of nutrients, exogenous nitrogen can be deleterious to the biological system. When the required nutrients are available to the plant, it is able to utilize the available N sources. Altering soil fertility by adding balanced amounts of minerals or organic sources of P, K and S could enhance the efficiency of N_2 fixation in common beans as well as its ability for utilizing other sources of N from soil and fertilizer.

Table 5. Effect of different levels of N, P, K, and S on the percentage and amount of nitrogen derived from air (% NDFA and ANDFA, respectively) and from soil (% NDFS and ANDFA respectively) by bean plants 90 days after germination

Fertility level	% NDFA	ANDFA (mg)	% NDFS	ANDFS (mg)
Low	51,9 B	131,0 C	27,8 A	69,4 B
5	65,7 a	122,4 ab	31, 7 a	58,5 ab
15	64,3 a	185,5 a	28,6 a	80,7 ab
60	42,9 b	128,3 ab	28,7 a	82,0 a
120	34,5 b	88,0 b	21,1 b	56,4 b
Medium	66,0 A	217,2 B	11,7 C	39,8 C
5	88,9 a	257,1 ab	9,5 b	28,4 b
15	87,3 a	337,4 a	8,7 b	33,8 b
60	66,8 b	187,1 b	11,6 ab	32,4 b
120	21,1 c	86,9 c	16,8 a	64,8 a
High	64,7 A	332,1 A	23,4 B	117,5 A
5	74,6 a	394,6 a	24,0 ab	126,8 a
15	70,7 a	246,7 b	25,1 a	86,5 b
60	57,0 b	312,1 ab	25,8 a	135,9 a
120	56,5 b	375,2 a	18,7 b	120,8 a
CV%	11	26	18	21

Means followed by the same letters do not differ statistically by Duncan Test ($p < 0.05$).
* Low = No P, K, and S addition.
Medium = 50, 63, and 10 mg kg^{-1} of P, K, and S, respectively.
High = 200, 256, and 40 mg kg^{-1} of P, K, and S, respectively.
N: 5, 15, 60 and 120 mg N kg^{-1}.

Acknowledgements

This contribution was supported by the U.S. National Academy of Sciences – BOSTID (BNF-4-86-53) and IAEA (311-D1-BRA-4249) research grants. The authors would like to thank Dr. Segundo C. Urquiagua, Marli de Fátima Fiore and Maria de Lourdes Liva for their support and assistance. S.M.T. acknowledges funding from Conselho Nacional de Pesquisa e Desenvolvimento (Brazil).

References

Bonetti R, Montanheiro M N S and Saito S M T 1984 The effects of phosphate and soil moisture on the nodulation and growth of *Phaseolus vulgaris* L. J. Agric. Sci. 103, 95–102.

Cassman K G, Whitney A S and Fox R L 1981 Phosphorus requirements of soybean and cowpea as affected by mode of N nutrition. Agron. J. 73, 17–22.

Danso S K, Bole J B and Zapata F 1983 A guide to the use of nitrogen-15 and radioisotopes in the studies of plant nutrition: Calculations and interpretations of data. IAEA-TECDOC-288. IAEA, Vienna, 153 p.

França G E, Bahia Filho A F C and Carvalho M M de. Influencia de magnesio, micronutrientes e calagem no desenvolvimento e fixação simbiotica de nitrogenio na soja perene var. Tinaroo (*Glycine weightii*) em solo de cerrado. Pesqui. Agropecu. Bras. 8, 197–202.

Franco A A and Day J M 1980 Effects of lime and molybdenum on nodulation and nitrogen fixation of *Phaseolus vulgaris* L. in acid soils of Brazil. Turrialba 30, 99–105.

Gates C T and Muller W J 1979 Nodule and plant development in the soybean, *Glycine max* (L) Merr.: Growth response to nitrogen, phosphorus and sulfur. Aust. J. Bot. 27, 203–215.

Gates C T and Wilson J R 1974 The interaction of nitrogen and phosphorus on the growth, nutrient status and nodulation of *Stylosanthes humilis* H.B.K. (Townsville Stylo). Plant and Soil 41, 325–333.

Graham P H 1981 Some problems of nodulation and symbiotic fixation in *Phaseolus vulgaris* L.: A review. Field Crops Res. 4, 93–112.

Graham P H and Rosas J C 1979 Phosphorus fertilization and symbiotic nitrogen fixation in common bean (*Phaseolus vulgaris* L.). Agron. J. 71, 925–927.

Hernandez B S and Focht D D 1985 Effects of phosphorus, calcium, and Hup^- and Hup^+ rhizobia on pigeon pea yields in an infertile tropical soil. Agron. J 77, 867–871.

Kungl G 1987 Acetylene reduction activity of *Glycine max* with response to inoculation, NPK application and soil environment. 9th Int. Symp. on Soil Biol. and Conserv. of the Biosphere, ed. J. Szegi, Akadémiai Kiadó, Budapest, Hungary, pp 265–272.

Lynd J G and Ansman T R 1989 Effects of P, Ca with four K levels on nodule histology, nitrogenase activity and improved 'spanco' peanut yields. J. Plant Nutr. 12, 65–84.

Lynd J G and Ansman T R 1990 Soil conditions with distinctive coralloid nodulation and nitrogen fixation of 'Mecca' alfalfa. J. Plant Nutr. 13, 77–94.

Martinez-Romero E, Segovia L, Mercante F M, Franco A A, Graham P H and Pardo M A 1991 *Rhizobium tropici*, a novel species nodulating *Phaseolus vulgaris* and *Leucaena* sp. trees. Int. J. Syst. Bacteriol. 41, 417–426.

Pereira P A A and Bliss F A 1987 Nitrogen fixation and plant growth of common bean (*Phaseolus vulgaris* L.) at different levels of phosphorus availability. Plant and Soil 104, 79–84.

Pereira P A A and Bliss F A 1989 Selection of common bean (*Phaseolus vulgaris* L.) for N_2 fixation at different levels of available phosphorus under field and environmentally-controlled conditions. Plant and Soil 115, 75–82.

Proksch G 1969 Routine analysis of ^{15}N in plant material by mass-spectrometry. Plant and Soil 31, 380–384.

Rosas J C and Bliss 1986 Improvement of the nitrogen fixation potential of common beans in Latin America. CEIBA 27, 245–259.

Ruschel A P and Reuszer H W 1973 Fatores que afetam a simbiose *Rhizobium phaseoli–Phaseolus vulgaris*. Pesqui. Agropecu. bras. 8, 287–292.

Ruschel A P and Saito S M T 1977 Efeito da inoculação de *Rhizobium* nitrogenio e matéria organica na fixaçao simbiotica do nitrogenio em feijão (*Phaseolus vulgaris* L.). Rev. Bras. Ci. Solo 1, 21–24.

Ruschel A P, Saito S M and Tulmann Neto A 1979a Efficiency of *Rhizobium* inoculation on *Phaseolus vulgaris*. I. Effects of nitrogen sources and plant variety. R. bras. Ci. Solo 3, 13–17.

Ruschel A P, Vose P B, Victoria R L and Salati E 1979b Comparison of isotope techniques and non-nodulating isolines to study the effect of ammonium fertilization on dinitrogen fixation in soybean, *Glycine max*. Plant and Soil 53, 513–525.

Saito S M T, Montanheiro M N, Victoria R L, Reichardt K 1984 The effects of N fertilizer and soil moisture on the nodulation and growth of *Phaseolus vulgaris* L. Physiol. Plant. 103, 87–93.

Saito S M T and Ruschel A P 1978 Influence of lime, phosphate and micronutrients on the natural nodulation of beans (*Phaseolus vulgaris* L.). Anais de ESALQ 35, 545–556.

Silva P M, Tsai S M and Bonetti R 1993 Response to inoculation and N fertilization for increased yield and biological nitrogen fixation of common bean (*Phaseolus vulgaris* L.). Plant and Soil 152, 123–130.

Sistachs E 1970 Efectos de la fertilización nitrogenada y la inoculación en el rendimento y contenido de nitrogeno del frijol negro (*Phaseolus vulgaris*). Rev. Cubana Ci. Agric., 4, 233–237.

Thomas R J and Hungria M 1988 Effect of K on N_2 fixation nitrogen transport and N harvest index of beans (*Phaseolus vulgaris* L.). J. Plant Nutr. 11, 175–188.

Vargas A A T and Graham P H 1989 Cultivar and pH effects on competition for nodule sites between isolates of *Rhizobium* in beans. Plant and Soil 117, 195–200.

Vernichenko L Y 1984 The effect of increasing levels of mineral nitrogen on the productivity of symbiotic nitrogen fixation. *In* Soil Biology and Conservation of the Biosphere. Ed. J Szegi Vol. 2, pp 487–493. Akadémiai Kiadó, Budapest.

Plant and Soil **152**: 139–143, 1993.

PLSO BNF15

Effect of different levels of mineral nitrogen on nodulation and N_2 fixation of two cultivars of common bean (*Phaseolus vulgaris* L.)

SABINE MÜLLER[1], P.A.A. PEREIRA[2] and P. MARTIN[1]
[1]*Institute for Plant Nutrition, University of Hohenheim, 70593 Stuttgart, Germany*;
[2]*EMBRAPA-CNPAF, Caixa Postal 179, 74000 Goiânia, Goias, Brazil*

Key words: difference method, N_2 fixation, nitrogen supply, nodulation

Abstract

An experiment was conducted under greenhouse conditions to evaluate the effect of mineral nitrogen on N_2 fixation of two cultivars of *Phaseolus vulgaris* L., Puebla 152 and Negro Argel. Nitrogen application was 0, 2.5, 12.5 and 25 mg N kg^{-1} of a vermiculite-sand-mixture at planting time. Shoot and root growth were elevated by nitrogen application at all growth stages. During vegetative growth (V 5) nodule dry weight and nitrogenase activity (acetylene reducing activity) per plant were reduced by nitrogen supply in both cultivars, but less in Negro Argel than in Puebla 152. At later stages nodulation in nitrogen-treated Puebla 152 did not differ from that in non-treated plants, whereas increased nodule number was found in Negro Argel at high nitrogen levels. The influence of mineral N on the total amount of nitrogen fixed in the two bean cultivars was only slightly different.

Introduction

Common bean (*Phaseolus vulgaris* L.) is considered to have poor symbiotic N_2 fixation potential compared to other legumes (Hardy et al., 1968; Piha and Munns, 1987). Field studies have shown that N fertilization increased yield substantially, indicating that the symbiotic nitrogen fixation is unable to provide enough nitrogen for maximum yield (Buttery et al., 1987; Huntingdon et al., 1986). Because of the weak performance of the symbiotic system in *Phaseolus vulgaris*, N fertilization is a common practice in most areas especially where growth conditions are sub-optimal.

In Brazil, fertilizer recommendations include applications of up to 20 kg N ha^{-1} at planting time alone, or together with a supplemental side-dressing of 40 kg N ha^{-1} during vegetative growth, depending on soil type, soil nutrient status and cropping history (Santana, 1985). It is thought that low N application may stimulate N_2 fixation by improving early plant growth until N_2 fixation provides adequate N for plant growth and development. On the other hand, fertilizer rates exceeding those exerting a "starter-N" effect, generally have reduced nodulation and N_2-fixation. In common bean the threshold for the depressing effect of mineral N seems to be low (Graham, 1981). Therefore, plants supporting a symbiotic system with a high tolerance against mineral N would be desirable. In the present study two cultivars of *Phaseolus vulgaris*, Puebla 152 and Negro Argel, were compared for their symbiotic performance and N_2-fixation in response to different amounts of mineral N applied at the time of sowing.

Material and methods

The experiment was conducted in the greenhouse of EMBRAPA-CNPAF in Goiânia, Brazil using the common bean cultivars, Puebla 152 and Negro Argel, both considered to have high N_2 fixation potential.

The nitrogen treatment were 0, 2.5, 12.5 and 25.0 mg N kg^{-1} of substrate applied as KNO_3 at planting. At each N level replicated pots with non-nodulating soybean isoline 'Harosoy' were included in the experiment. Two plants were grown in each 7 kg pot containing a N-free 2:1 vermiculite:sand mixture. 100 mL per pot of a N-free modified Summerfield's nutrient solution (Summerfield et al., 1977) was applied every two days and distilled water was supplied to maintain optimum moisture of the substrate. All the pots received 0.4 mL suspension with 10^8 cell of the *Rhizobium leguminosarum* biovar *phaseoli* strain CIAT 899 at planting.

Plant samples were taken at the V_5, R_1, R_7 growth stages to evaluate nodule number, nodule dry weight, shoot dry weight, shoot N content and nitrogenase activity by the acetylene reduction method (Hardy et al., 1968). Non-nodulating soybeans were harvested at R_7 stage to estimate the amount of N_2 fixed by common bean using the N difference method (Boddey, 1987). The experimental design was a randomized complete block design. Analysis of variance was done by MSTAT statistical programme and the Duncan's Multiple Range test was used for mean separation.

Results and discussion

Shoot dry weight and, even more, root dry weight, increased at all sampling dates when nitrogen was applied, with Negro Argel showing a greater response than Puebla 152 to mineral N (Table 1). Since N was applied at seeding time, available N decreased gradually. This is reflected by a strong effect on plant growth at early stages, and a smaller influence at later stages. Plant dry weight of Puebla was higher than that of Negro Argel, but percentage of N in the shoot of Puebla was less. Therefore, no cultivar differences in nitrogen accumulation per plant could be found (Table 1).

Nodule number of Puebla was mostly higher than that of Negro Argel, except at levels of high nitrogen application, but individual nodule dry weight was generally less (Table 2). The mineral N decreased nodule size, nodule dry weight and nitrogenase activity (acetylene reducing activity)

Table 1. Shoot and root dry weight, nitrogen content of shoot and N_2 fixed (N difference method) of two cultivars of *Phaseolus vulgaris* L., Puebla 152 (P.) and Negro Argel (N.A.), grown with two levels of mineral nitrogen at different growth stages (V5, R1, R7). Values in parenthesis: Percent N_2 fixed of total N

mg N kg^{-1} substrate	V_5		R_1		R_7	
	P	N.A.	P	N.A.	P	N.A.
Shoot dry weight (g $plant^{-1}$)						
0	1.38 c	0.92 d	3.83 b	2.93 c	11.61 bc	10.31 c
25	1.99 a	1.66 b	5.32 a	4.77 a	13.55 a	13.28 ab
Root dry weight (g $plant^{-1}$)						
0	0.20 b	0.12 c	0.39 b	0.25 b	0.49 bc	0.38 c
25	0.35 a	0.34 a	0.78 a	0.68 a	0.95 a	0.62 b
Nitrogen content in shoot (mg $plant^{-1}$)						
0	33.7 b	29.5 a	99.4 b	95.0 b	221.1 b	205.9 b
25	51.3 a	44.9 a	117.9 ab	140.3 a	228.7 ab	281.3 a
Nitrogen fixed (mg $plant^{-1}$)						
0					212 a (96)	197 a (95)
25					189 a (82)	242 a (86)

Means with different letters indicate significant differences ($p < 0.05$) among cultivars and N-treatments at one sampling date.

Table 2. Nodule number, nodule size and specific acetylene reducing activity of two cultivars of *Phaseolus vulgaris* L., Puebla 152 (P.) and Negro Argel (N.A.), grown with four levels of mineral nitrogen at different growth stages (V_5, R_1, R_7)

mg N kg^{-1} substrate	V_5		R_1		R_7	
	P	N.A.	P	N.A.	P	N.A.
Nodule number per plant						
0	198 a	110 c	405 ab	282 b	559 ab	347 b
2.5	186 ab	115 c	398 ab	367 ab	550 b	486 ab
12.5	155 abc	126 bc	335 b	393 ab	543 ab	434 ab
25.0	103 c	108 c	339 a	472 a	575 ab	610 a
Nodule D.W. (mg D.W. $nodule^{-1}$)						
0	0.52 b	0.86 a	0.61 c	0.88 a	0.65 ab	1.00 ab
2.5	0.55 b	0.85 a	0.61 c	0.72 abc	0.67 ab	0.64 ab
12.5	0.36 bc	0.73 a	0.72 abc	0.82 ab	0.70 ab	0.90 ab
25.0	0.27 c	0.51 b	0.58 c	0.67 bc	0.64 ab	0.55 b
Specific acetylene reducing activity (μmol C_2H_4 $g^{-1}h^{-1}$ nodule D.W.)						
0	18.0 ab	16.7 b	7.36 a	8.29 a	1.14 bc	1.18 bc
2.5	20.4 ab	18.0 ab	8.77 a	8.39 a	1.07 c	0.93 c
12.5	21.0 a	19.5 ab	9.49 a	8.25 a	1.25 bc	1.39 abc
25.0	17.2 ab	20.2 ab	8.86 a	8.42 a	1.60 ab	1.82 a

Means with different letters indicate significant differences ($p \leq 0.05$) among cultivars and N-treatments at one sampling date.

per plant during the growth phase until V5 (Fig. 1, 2). The effect was stronger in Puebla. A positive response was found in Negro Argel for nodule dry weight per plant at higher N levels at R_1 stage (Fig. 1), resulting from an increased nodule number at later stages (Table 2). Mineral nitrogen also caused elevated specific acetylene reducing activity in nodules at R_7 growth stage (Table 2). These results confirm positive effects of starter-N application on nodulation as described previously (Park and Buttery, 1989; Srivastava and Ormond, 1986).

A non-nodulating soybean was included as a reference plant in this study to differentiate between N taken up from the substrate and N_2 fixed. Nitrogen fixation contributed strongly to

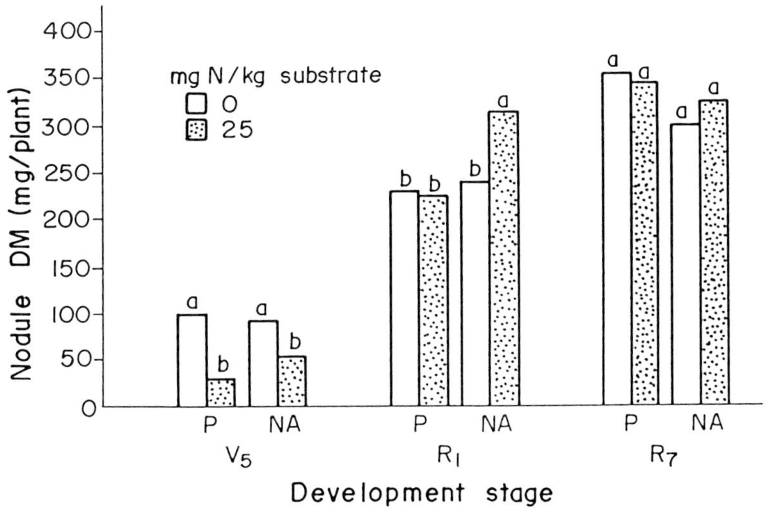

Fig. 1. Nodule dry weight of two cultivars of *Phaseolus vulgaris*, Puebla 152 (P) and Negro Argel (NA) grown with two levels of mineral nitrogen at different growth stages (V_5, R_1, R_7). Means with different letters indicate significant differences ($p \leq 0.05$) among cultivars and N-treatments at one sampling date.

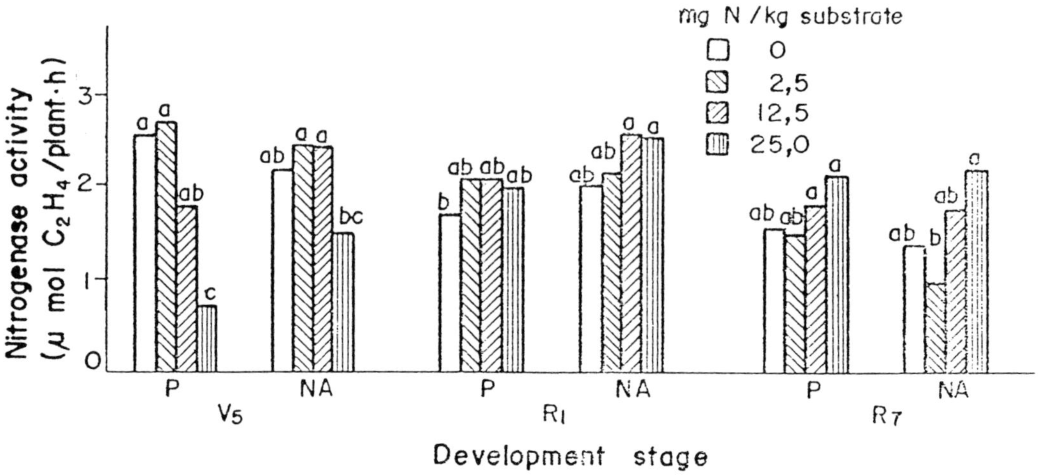

Fig. 2. Acetylene reducing activity of two cultivars of *Phaseolus vulgaris*, Puebla 152 (P) and Negro Argel (NA) grown with four levels of mineral nitrogen at different growth stages (V_5, R_1, R_7). Means with different letters indicate significant differences ($p \leq 0.05$) among cultivars and N-treatments at one sampling date.

the nitrogen accumulation of both bean cultivars at all nitrogen levels (Table 1). Although at R_1 Negro Argel had more nodule dry weight and nitrogenase activity on a per plant basis than Puebla, there was no significant difference between the two cultivars in the amount of N from fixation at R_7. This might result from a stronger increase of nodule dry weight per plant in Puebla than in Negro Argel after R_1 growth stage (Fig. 1). It is suggested, that the greater part of fixation takes place after flowering as previously reported (Zapata et al., 1987a,b). In spite of a decrease in percentage of nitrogen from atmosphere with increasing application of mineral nitrogen, the amount of N_2 fixed was not influenced significantly.

Studies with common bean and soybean have indicated that the sensitivity of N_2 fixation to combined N also depends on rhizobial strain (Awonaike et al., 1980; Senaratne et al., 1987). In pea (*Pisum sativum* L.) isolates of *Rhizobium* varied in effectiveness in the presence of combined N, but differences in symbiotic response were small and it was concluded that efforts to improve nitrogen fixation in the presence of combined nitrogen should be directed towards host cultivars rather than *Rhizobium* strains (Nelson, 1987). In common bean currently much effort is being made to identify and characterize genotypes with superior nodulation and N_2 fixation (Pereira et al., 1989; Rennie and Kemp, 1983, Ruschel et al., 1982; Wolyn et al., 1989). The results of the present study provide evidence for the variability of bean cultivars grown with different amounts of mineral N. It seems worthwhile to continue identifying lines that can use both sources of N effectively, to obtain parents for breeding programmes to improve N_2 fixation in beans.

Acknowledgement

The study was supported in part by LGT-G/DAAD (Germany).

References

Awonaike, K O, Lea P J, Day J M, Roughley R J and Miflin B J 1980 Effects of combined nitrogen on nodulation and growth of *Phaseolus vulgaris*. Expl. Agric. 16, 303–311.

Boddey R M 1987 Methods for quantification of nitrogen fixation associated with gramineae. Crit. Rev. Pl. Sci. Vol. 6, 209–266.

Buttery B R, Park S J and Findlay W I 1987 Growth and yield of white bean (*Phaseolus vulgaris*) in response to nitrogen, phosphorus and potassium fertilizer and to inoculation with *Rhizobium*. Can. J. Plant Sci. 67, 425–432.

Graham P H 1981 Some problems of nodulation and symbiotic nitrogen fixation in *Phaseolus vulgaris*: A review. Field Crops Res. 4, 93–112.

Hardy R W F, Holsten R D, Jackson E K and Burns R C 1968 The acetylene-ethylene assay for N_2-fixation: Laboratory and field evaluation. Plant Physiol. 43, 1185–1207.

Huntington T G, Smith M S, Thomas G S, Blevins R L and Perez A 1986 Response of *Phaseolus vulgaris* to inoculation with *Rhizobium phaseoli* under two tillage systems in the Dominican Republic. Plant and Soil 95, 77 – 85.

Nelson L M 1987 Response of *Rhizobium leguminosarum* isolates to different forms of inorganic nitrogen during nodule development in pea (*Pisum sativum* L.). Soil Biol. Biochem. 19, 759–763.

Park S J and Buttery B R 1989 Identification and characterization of common bean (*Phaseolus vulgaris* L.) lines well nodulated in the presence of high nitrate. Plant and Soil 119, 237–244.

Pereira P A A, Burris R H and Bliss F A 1989 ^{15}N determined dinitrogen fixation potential of genetically diverse bean lines (*Phaseolus vulgaris* L.). Plant and Soil 120, 171–179.

Piha M I and Munns D N 1987 Nitrogen fixation potential of beans (*Phaseolus vulgaris* L.) compared with other grain legumes under controlled conditions. Plant and Soil 98, 169–182.

Rennie R J and Kemp G A 1983 N_2-fixation in field beans quantified by ^{15}N isotope dilution. II. Effect of cultivars of beans. Agron. J. 75, 645–649.

Ruschel A P, Vose P B, Matsui E, Victoria R L and Tsai Saito S M 1982 Field evaluation of N_2-fixation and N-utilization by *Phaseolus* bean varieties determined by ^{15}N isotope dilution. Plant and Soil 65, 397–407.

Santana M B M 1985 Ed. Adubacao nitrogenada no Brasil. CEPLAC Ilheus, Bahia, Brazil.

Senaratne R, Amornpimol C and Hardarson G 1987 Effect of combined nitrogen on nitrogen fixation of soybean (*Glycine max*. L. Merrill.) as affected by cultivar and rhizobial strain. Plant and Soil 102, 45–50.

Srivastava H S and Ormond D P 1986 Effects of nitrogen dioxide and nitrate nutrition on nodulation, nitrogenase activity, growth and nitrogen content of bean plants. Plant Physiol. 81, 737–741.

Summerfield R J, Dart P J, Huxley P A, Eaglesham A R J, Minchin F R and Day J M 1977 Nitrogen nutrition of cowpea (*Vigna unguiculata*). I. Effects of applied nitrogen and symbiotic nitrogen fixation on growth and seed yield. Expl. Agric. 13, 129–142.

Wolyn D J, Attewell J, Ludden P W and Bliss F A 1989 Indirect measures of N_2-fixation in common bean (*Phaseolus vulgaris* L.) under field conditions: The role of lateral root nodules. Plant and Soil 113, 181–187.

Zapata F, Danso S K A, Hardarson G and Fried M 1987a Time course of nitrogen fixation in field-grown soybean using nitrogen-15 methodology. Agron. J. 79, 172–176.

Zapata F, Hardarson G, Danso S K A and Fried M 1987 b Nitrogen fixation and translocation in field-grown fababean. Agron. J. 79, 505–509.

Plant and Soil **152**: 145–150, 1993.

PLSO BNF16

Effect of fungicides on survival of *Rhizobium* on seeds and the nodulation of bean (*Phaseolus vulgaris* L.)

M.L.G. RAMOS and W.Q. RIBEIRO JR.
Embresa Goiana de Pesquisa Agropecuária, C.P. 49, Goiânia, GO, Brazil, 74.130

Key words: common bean, fungicides, nodulation, *Phaseolus vulgaris* L., *Rhizobium*, survival

Abstract

The survival of *Rhizobium leguminosarum* biovar *phaseoli* on seeds of bean was tested, using the cultivar Carioca. The seeds were treated seven days before inoculation with Benlate, Vitavax, Banrot, Difolatan or Ridomil fungicides. The rhizobial strains used were: CIAT 899, CPAC 1135 and CIAT 652. Strain CIAT 899 showed greater survival on the seed with fungicide than the other strains. Two hours after the contact with fungicides strains CIAT 652 and CPAC 1135 had significantly lower numbers of rhizobia than the treatment without fungicide. The Benlate and Banrot fungicides had the greatest effect on survival of rhizobial strains. There was a drastic mortality of the two strains, CIAT 652 and CPAC 1135, on seeds treated with Benlate and Ridomil. Under field conditions, granular inoculation produced fewer nodules, but a similar total nodule weight as seed inoculation. Serological tests (ELISA) showed that seed treatment with Benlate in connection with seed inoculation reduced drastically the occurrence of inoculated strains in nodules, while the same fungicide treatment and inoculation applied in the seed furrow did not affect the survival of the inoculated strain.

Introduction

Bean seeds are often treated with fungicides, like Benlate, Banrot or Ridomil to prevent soil born diseases caused by *Rhizoctonia*, *Fusarium* and *Pythium*. There are studies showing the incompatibility of some fungicides with *Rhizobium leguminosarum* biovar *phaseoli* affectiving its survival on seeds, and subsequently nodulation and nitrogen fixation (Graham et al., 1980; Kecskes and Vincent, 1973). However, there are no data available for other recommended fungicides on their compatibility with *Rhizobium* (De-Polli et al., 1986). The placement of the inoculants, e.g., the use of granular inoculants, applied in the planting furrows without direct contact with the fungicide, may solve the incompatibility problem between strains and fungicides (Chamber and Montes, 1982).

The objectives of the present study were: (i) to study the effects of fungicides on the survival of *Rhizobium* on common bean seeds and, (ii) to evaluate the early nodulation through two inoculation methods.

Materials and methods

Rhizobial survival on seed

The strains CIAT 899, CPAC 1135 and CIAT 652, of *Rhizobium leguminosarum* biovar *phaseoli*, were inoculated into sterilized peat, originating from Australia. After mixing the peat with the rhizobial suspension, the inoculants were maintained at room temperature for 7 days and then stored at 10°C. The inoculants with strains CIAT 899, CPAC 1135 and CIAT 652 contained 3.7×10^9, 1.1×10^9 and 2.9×10^9 cells per gram of peat, respectively.

Seeds of Carioca, a common bean cultivar grown widely in Brazil, were disinfected for 3 minutes in alcohol, 3 minutes in sodium hypochloride (3%), and then washed 6 times with sterilized distilled water, prior to fungicide treatment.

The fungicides tested in the first experiment were: Ridomil 25% (alfa-(2 chlorophenyl)- alfa-4 (chlorophenyl)-5-pyrimidinemethanol), Benlate 50% (methyl-1-butylcarbamoil)-2-benzimidazol-carbamate), Vitavax 75% (5,6 dihydro-2 methyl-1.4- oxatin-3- carbaxanilide, Banrot 40% (3-(2-methylpiperidino)-propyl - 3,4-dichlorobenzoate) and Difolatan 80% (cis-N-(1,1,2,2- tetrachloro-ethyltrio)-4-cyclohexane-1,2-dicarboximide). Each fungicide was applied at the recommended doses of 2.5 g kg^{-1} of seeds, and added as a triton (2%) solution.

Seven days after the fungicide treatment, 10 g of seeds were inoculated with a paste of 0.1 g inoculant in 25% sucrose solution. At 2, 24 and 48 h after inoculation, the number of viable cells were counted by the dilution method in Petri dishes containing yeast-extract mannitol-agar (YEM) (Vincent, 1970). Three replications were performed. For each dilution series, seeds were incubated after inoculation.

A second experiment had similar treatments with Benlate and Ridomil fungicides and rhizobial inoculants using the same doses of inoculants and fungicides as the previous experiment. The strains tested were CIAT 899 and CPAC 1135. The survival of *Rhizobium* on the seeds was also evaluated at 2, 24 and 48 h after inoculation.

Field evaluation of nodulation

A field trial was carried out to test nodulation of plants treated with fungicides. The cultivar used was Carioca, and the seeds were inoculated with the rhizobial strain CIAT 652.

The following treatments were tested: (1) granular inoculant applied in the furrow; (2) granular inoculant and Benlate applied simultaneously in furrows; (3) seed inoculation; (4) seed inoculation + Benlate application 7 days before inoculation; (5) seed inoculation and simultaneous Benlate application; (6) no inoculation. The inoculation rate on the seeds was the same as in the previous experiments. Granular inoculant was applied in the furrow at a rate of 20 kg ha^{-1} (95% sand + 5% inoculant), resulting in approximately 3×10^5 cells m^{-1} of row. Benlate application was made the same way as in the previous experiments. The treatments 3, 4 and 5 were sowed 2 hours after inoculation. All treatments with no rhizobial inoculation of the seeds were treated with sterile peat + sucrose 25% solution. Plots were $2m^2$ with 20 plants after thinning. The experiment was conducted as a randomized complete block design.

Twenty-two days after sowing, 10 plants per plot were harvested at random to determine nodule weight and number and shoot dry weight. Twenty nodules from 5 plants were sampled to determine the occurrence of the inoculated strains by the ELISA technique (Clark and Adams, 1977).

Results and discussion

Rhizobial survival on seed

Two hours after inoculation with strain CIAT 899, Difolatan and Vitavax had no significant influence on rhizobia survival. However, the other fungicides reduced rhizobia survival compared to the untreated control. All chemicals showed deleterious effects on rhizobial survival compared to the control 24 hours after fungicides treatments. The same tendency was observed when inoculant and fungicides were in contact (Fig. 1).

Benlate, Banrot and Difolatan had the greatest influence on the survival of strain CPAC 1135 after 2, 24 and 48 hours (Fig. 1). Benlate and Banrot had the most deleterious effects on all rhizobial strains with as little as 24 hours contact between rhizobia and fungicides. Survival of strain CIAT 652 was reduced by all fungicides compared to the control 2 hours after incubation. For this strain, Benlate and Difalatan affected survival, while Ridomil had the least deleterious effect on rhizobial survival 24 and 48 hours after the fungicide treatments (Fig. 1). Even though the numbers of rhizobial cells of the three strains that were on the seeds initially were different for the no-fungicide treatment

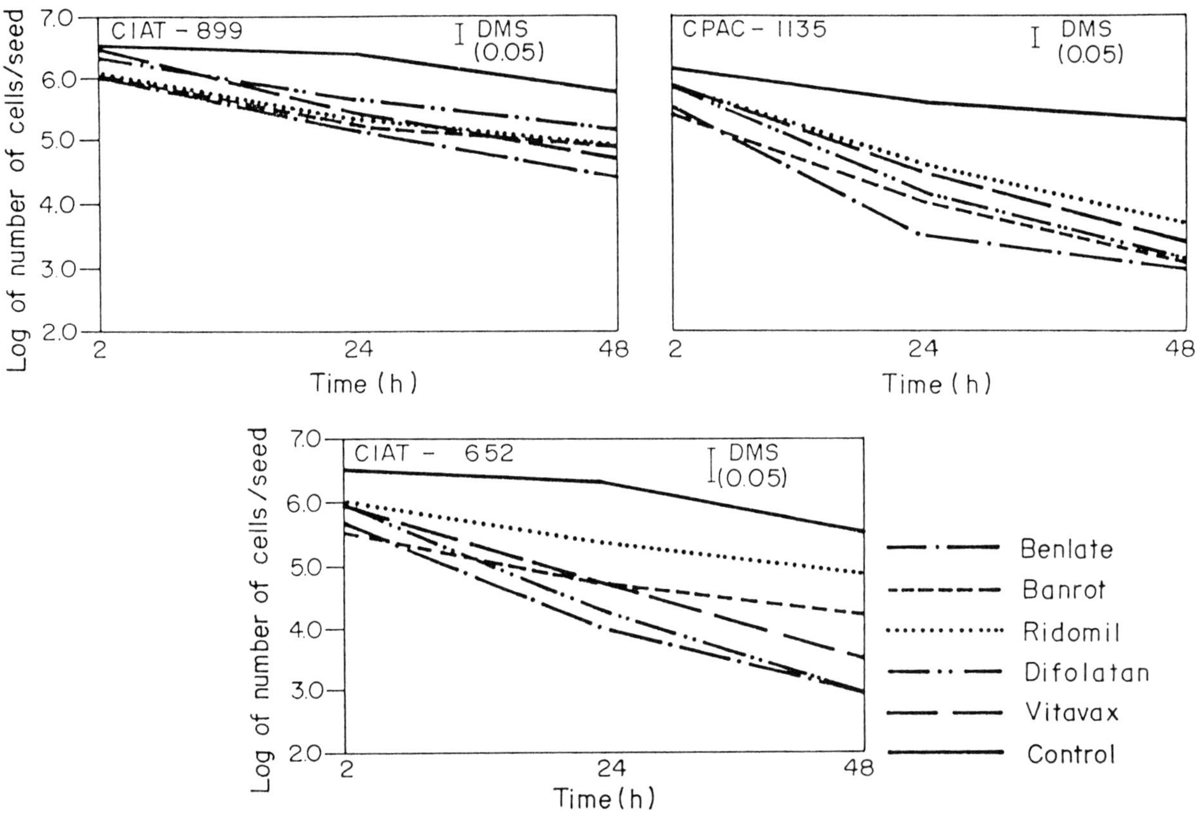

Fig. 1. Survival of *Rhizobium* strains on seed treated 7 days before inoculation with the fungicides Benlate, Ridomil, Difolatan, Vitavax and Banrot, as compared to the control (without fungicide).

(Fig. 1), strain CIAT 899 survived longer in the presence of fungicide than the other strains. Two hours following the fungicide treatment, survival of strains CIAT 652 and CPAC 1135 was affected more than strain CIAT 899. Benlate and Benrot also influenced rhizobial survival of the three strains more than the other fungicides.

Rhizobium species and strains have been reported to be affected differently by fungicides and insecticides (Diatloff, 1970; Graham et al., 1980; Kapusta and Rouwenhourst, 1973, Lehri, 1976). In the present study, Vitavax showed a strong deleterious effect on survival of all three strains even after 24 hours, contrary to the finding of Curley and Burton (1975). Strains have been found to metabolize fungicides to other non-toxic compounds (Odeyemy and Alexander, 1977a).

Ridomil and Benlate caused a drastic reduction in cell survival of CIAT 899 on the seeds with simultaneous seed inoculation and fungicide treatment. This was nearly 3 logarithm units for Ridomil and 1.5 for Benlate in only 2 hours of fungicide contact (Fig. 2). Benlate had a slower deleterious effect than Ridomil on rhizobia survival.

The simultaneous seed inoculation and fungicide treatment greatly decreased the survival of the strains on the seeds compared to rhizobial inoculation 7 days after seed treatment with Benlate and Ridmil (Figs. 1, 2). Lopes and Portugal (1986) also showed that the simultaneous fungicide and rhizobial inoculation treatment affected *Rhizobium* survival on seeds more than did seed treatment and inoculation done 10 days apart.

Field evaluation of nodulation

Granular inoculant placed in the furrow resulted in fewer nodules, but in similar total weight

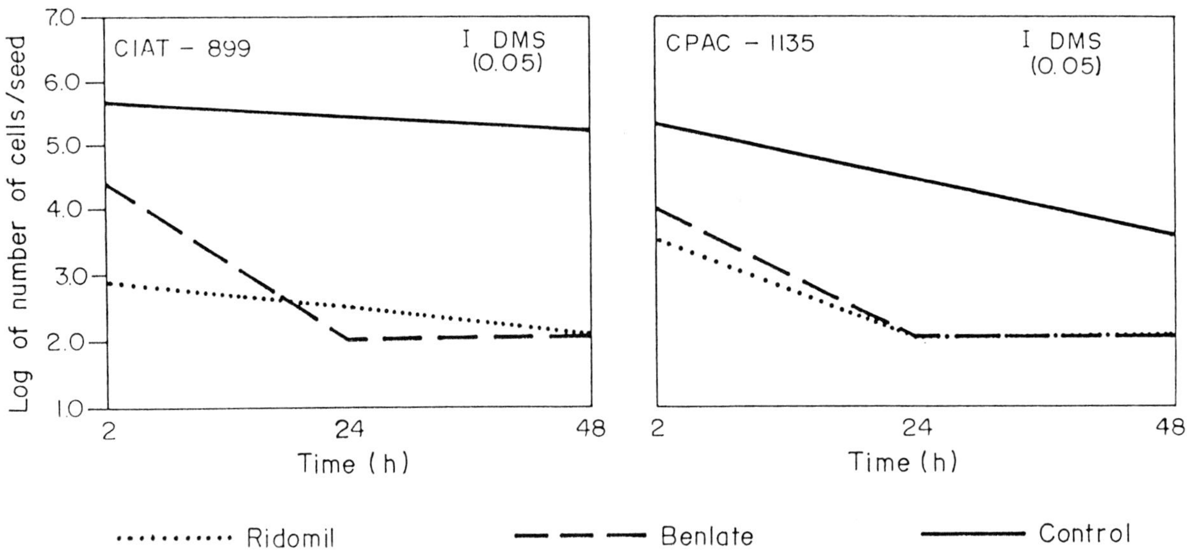

Fig. 2. Survival of *Rhizobium* strains inoculated and simultaneously treated with Benlate and Ridomil as compared to the control without fungicide.

compared to the seed inoculation treatment (Table 1). Similar results were observed for soybean (Chamber and Montes, 1982). The treatments with fungicides and seed inoculation resulted in greater nodule weight. Seed inoculation produced more nodules than treatments using granular inoculant (Table 1). Previous studies have shown that seed treatment with fungicides can cause an increase (Lopes and Portugal, 1986), a reduction (Jones and Giddens, 1984; Rennie and Dubetz, 1984) or no effect (Kucey and Bonetti, 1988) on nodulation. On the other hand, nutrient deficiency of molybdenum and cobalt, may decrease nodule number and weight (Robson, 1973), complicating the identification of the deleterious effect of the pesticide. No significant differences in shoot dry weight were observed among treatments, 20 days after planting (Table 1).

The fungicide seed treatments affected the occurrence of the inoculated strain in the nodules (Table 2). The use of granular inoculant and Benlate did not reduce inoculant strain occurrence, and this treatment did not differ significantly from application of granular inoculant without fungicide (Table 2). The fungicide application on seeds increased nodulation (Table 1), but reduced the occurrence of the inoculated strain in nodules (Table 2). Some of the indigenous rhizobia strains were of sero group that cross reacted with strain CIAT 652.

Studies of compatibility between fungicides

Table 1. Effect of Benlate and inoculation methods on nodule number, nodule dry weight and shoot dry weight under field conditions

Treatments	Nodule weight (mg/plant^{-1})	Nodule number (No/plant^{-1})	Shoot weight (g/plant^{-1})
Granular inoculant	37.2 bc[1]	37.3 c	0.7 ns
Granular inoculant + Benlate	23.9 c	42.4 bc	0.7 ns
Seed inoculation	24.7 c	59.0 ab	0.6 ns
Seed inoculation + Benlate[2]	42.7 ab	65.8 a	0.7 ns
Seed inoculation + Benlate[3]	55.2 a	55.4 abc	0.7 ns
No inoculation	20.5 c	66.3 a	0.6 ns

[1] The numbers followed by the same letter in the columns denotes no significance by Duncan test ($p < 0.05$); ns = non-significant.
[2] Inoculation 7 days after Benlate treatment.
[3] Simultaneous inoculation and Benlate treatment.

Table 2. Effect of Benlate and inoculation methods on percentage of nodules formed by the inoculated strain under field conditions

Granular inoculant	61.2 a[1]
Granular inoculant + Benlate	50. a
Seed inoculation	50.8 a
Seed inoculation + Benlate[2]	21.2 b
Seed inoculation + Benlate[3]	23.5 b
No inoculation	16.5 b

[1] The numbers followed by the same letter in the columns denote no significance by the Duncan test ($p < 0.05$) from the data transformed to log x.
[2] Inoculation 7 days after Benlate treatment.
[3] Simultaneous inoculation and Benlate treatment.

and *Rhizobium* (Diatloff, 1970; Lopes and Portugal, 1986; Rennie and Dubetz, 1984), have been conducted in soils with large *Rhizobium* populations (Chamber and Montes, 1982; Fisher and Haynes, 1981; Mallik and Tesfai, 1985), which may explain the contradicting results in the literature. The present study was conducted in a soil with a large native *Rhizobium* population. Serological identification of strains in nodules showed a strong reduction in the occurrence of the inoculated strain compared to the control at 22 days after sowing when the seeds were treated with fungicides and seeds were inoculated (Table 2). Therefore, the drastic reduction in the survival of CIAT 652 in the field (Table 2) was due to contact with Benlate, similar to results observed in the laboratory (Fig. 1).

The use of rhizobial strains naturally tolerant to fungicides, can increase nodulation and maintain the survival of the inoculated strain on the seeds (Lennox and Alexander, 1981). Another possibility would be to produce mutants tolerant to fungicides (Odeyemi and Alexander, 1977b), although mutant strains often are less efficient than the original strain (Rai, 1985). Application of granular inoculant is not a common practice for beans, but could be useful when seeds are treated with fungicides. Therefore, a method of inoculation which prevents contact between rhizobia and fungicides, could be an alternative to seed inoculation. According to our results, seeds should not be simultaneously inoculated with rhizobia and treated with fungicides because this practice increases the rate of mortality of the inoculated strains on seeds compared to seed treatment with fungicides and inoculation seven days later.

Acknowledgements

We thank Drs. Joao Gaspar Farias and Pedro A.A. Pereira for valuable suggestions. This work was conducted at the International Center for Tropical Agriculture (CIAT) in Colombia when the authors were on study leave in the laboratory of J A Kipe-Nolt.

References

Chamber M A and Montes F J (1982) Effects of some seed disinfectants and methods of rhizobial inoculation on soybean (*Glycine max* L. Merrill.). Plant and Soil 66, 353–360.

Clark M F and Adams A N 1977 Characteristics of the microplate method of enzyme-linked immunosorbent assay (ELISA) for the detection of plant viruses. J. Gen. Virol. 34, 475–483.

Curley T and Burton J C 1975 Compatibility of *Rhizobium japonicum* with chemical seed protectants. Agron. J. 67, 807–808.

De-Polli H, Souto S M and Franco A A 1986 Compatibilidade de agrotoxicos com *Rhizobium* spp. e a simbiose das leguminosas. EMBRAPA-UAPNPBS, Seropédica, RJ. 75p.

Diatloff A 1970 The effects of some pesticides on root nodule bacteria and subsequent nodulation. Aust. J. Exp. Agric. Anim. Husb. 10, 562–567.

Fisher D J and Haynes A L 1981 Effects of some fungicides used against cereal pathogens on the growth of *Rhizobium trifolii* and its capacity to fix nitrogen in white clover. Ann. Appl. Biol. 98, 101–107.

Graham P H, Ocampo G, Ruiz L D Duque A 1980 Survival of *Rhizobium phaseoli* in contact with chemical seed protectants. Agron. J. 72, 625–627.

Jones R and Giddens J 1984 Introduction of effective N_2-fixing rhizobial strains into the soybean plant by use of fungicide resistance. Agron. J. 76, 599–603.

Kapusta G and Rouwnehorst D L 1973 Interaction of selection pesticides and *Rhizobium japonicum* in pure culture and under field conditions. Agron. J. 65, 112–115.

Kecskes M and Vicent J M 1973 Compatibility of fungicide treatment and *Rhizobium* inoculation of vetch seed. Acta Agron. Acad. Sci. Hung. 22, 1–2.

Kucey R M N and Bonetti R 1988 Effect of vesicular-arbuscular mycorrhizal fungi and captan on growth and N_2

fixation by *Rhizobium*-inoculated field beans. Can. Soil Sci. 68, 143–149.

Lehri L K 1976 "In vitro" inhibitory effect of some fungicides on *Rhizobium* spp. and *Azotobacter chroococcum*. J. Ind. Soc. Soil Sci. 24, 214–261.

Lennox L B and Alexander M 1981 Fungicide enhancement of nitrogen fixation and colonization of *Phaseolus vulgaris* by *Rhizobium phaseoli*. Appl. Environ. Microbiol. 41, 404–411.

Lopes E S and Portugal E P 1986 Compatibilidade entre o trataento de sementes d amendoim comfungicidas, sobrevivencia de *Rhizobium* e nodulacao. Bragantia 45, 293–302.

Mallik M A B and Tesfai K 1985 Pesticidal effect on soybean-rhizobia symbiosis. Plant and Soil 85, 33–41.

Odeyemy O and Alexander M 1977a Resistance of *Rhizobium* strains to phygon, spergon and thiran. Appl. Environ. Microibiol. 33, 784–790.

Odeyemi O and Alexander M 1977b Use of fungicide-resistant rhizobia *Rhizobium phaseoli*, *Rhizobium meliloti* for legume inoculation. Soil Biol. Biochem. 9, 247–251.

Rai R 1985 Competitiveness and effectiveness of carbedazin-resistant mutant and indigenous *Rhizobium* strains of chickpea (*Cicer arietnum* L.) in calcarious soil. J. Agric. Sci. 104–341–344.

Rennie R J and Dubetz S 1984 Effect of fungicides and herbicides on nodulation and N_2 fixation in soybean fields lacking indigenous *Rhizobium japonicum*. Agron. J. 76, 451–454.

Robson A D 1973 Mineral Nutrition. *In* Nitrogen Fixation Vol 3. Legumes. Ed. W J Broughton. pp. 36–55.

Saito S M T 1982 Avaliaco em campo da capacidade fixacao simbiotica de estirpes de Rhizobium phaseoli. Pesq. Agropec. brasileira 17, 999–1006.

Vincent J M 1970 A Manual for the practical Study of Root Nodule Bacteria. Blackwell Scientific, Oxford 164 p.

Plant and Soil **152**: 151–155, 1993.

PLSO BNF 17

Effects of water stress on N_2 fixation and grain yield of *Phaseolus vulgaris* L.

J.J. PEÑA-CABRIALES and J.Z. CASTELLANOS
Centro de Investigacion de Estudios Avanzados del IPN, Unidad Irapuato, Apartado Postal 629, 36500 Irapuato, Gto., Mexico

Key words: earliness, harvest index, nodulation

Abstract

Most dry bean production in Mexico is under non-irrigated conditions in the semi-arid highlands. One of the most limiting factors is insufficient moisture during the reproductive stage and sometimes during the vegetative stage. The objective of this experiment was to study the effect of drought on nodulation, N_2 fixation and grain yield of beans. The cultivars evaluated were: Flor de Mayo Bajio, Bayocel, Bat-477 and Honduras-35. The treatments were water stress treatments during vegetative or reproductive stage and a control of minimal stress. To measure N_2 fixation, ^{15}N-labelled fertilizer was used. Data for soil moisture, nodule number and nodule dry weight, as well as, straw and grain yield and total N were taken. The results indicated that water stress during vegetative stage depressed nodulation temporarily, but after watering regularly plants not only recovered completely but were nodulated better than the control. Water stress during the reproductive stage depressed nodulation and after watering there was no recovery. Grain yield was not affected by water stress during vegetative stage but it was reduced when water stress was imposed during the reproductive stage. The percentage of N derived from fixation varied among cultivars but was not affected by water stress treatment. The highest N_2 fixation occurred in Bayocel and Bat-477 and the lowest in Honduras-35 and F.M. Bajio, although the amounts were not as low as in some other reports.

Introduction

Eighty-five percent of the dry bean production area in Mexico is non-irrigated and most of this suffers drought conditions sometime during the growth of the plant. In the semi-arid highlands of Mexico, drought occurs regularly during the reproductive stage (Ibarra et al., 1987), although it may also occur during the vegetative stage. There is little information in the literature regarding the effect of drought on N_2 fixation of common bean (*Phaseolus vulgaris* L.), but studies in other grain legumes suggest that N_2 fixation is sensitive to drought and even more so than NO_3^- reduction (Sprent, 1982).

The stage of the crop during which drought occurs has an important effect on nodulation and N_2 fixation. An extended period of stress during the vegetative stage retards nodulation and depresses N_2 fixation (Abdel Ghaffar et al., 1982; Saito et al., 1984; Zablotowicz et al., 1981). Once the nodules are established, desiccation results in decreased N_2 fixation (Pankurst and Sprent, 1975), and it drops to near zero when nodule moisture content is reduced to 50 or 55%. The loss of nodule activity is reversible – if the nodules lose no more than 20% of the maximum fresh weight (Sprent, 1971). The effect of drought on nodule activity occurs primarily as a result in increasing of resistance to oxygen diffusion to the bacteroids (Durand et al., 1987). The objective of this experiment was to study the effect of drought during two growth stages in four cultivars of dry beans on nodulation, N_2 fixation and grain yield.

Materials and methods

This field experiment was carried out at the Agricultural Experimental Station in Celaya, Gto, Mexico on a Typic Pellusterts soil (pH 7.7, 1.2% organic matter and clay in texture).

Four cultivars of dry bean were studied, i.e., Flor de Mayo Bajio, Bayocel, Bat-477 and Honduras-35, and millet was included as a non-fixing reference crop. The bean seeds were inoculated by a peat base inoculant at the time of planting by a mixture of two strains (Guanajuato 2.1 and 12.1) of *Rhizobium leguminosarum* biovar. *phaseoli* prepared by the microbial ecology laboratory at CINVESTAV. The experiment was planted on March 7th, 1990. The bean cultivars were randomized within the large plots. The large plots corresponded to water stress treatments and the small plots referred to the different bean cultivars. Adjacent to the bean growing area, millet was cultivated as a reference crop. In all cases sub-plots had six rows, each 6 m long and 0.45 m between rows, with planting densities of 220,000 plant ha^{-1}. Each treatment was replicated four times. Immediately after sowing 10 kg N ha^{-1} as ammonium sulphate with 10% ^{15}N atom excess was applied to the 1.1 m^2 micro-plots. The remaining area received unlabelled ammonium sulphate at the same rate. All plots were fertilized with 60 kg P_2O_5 ha^{-1}.

Two water stress treatments were imposed: (a) during vegetative stage and (b) during reproductive stage, and a control with no water stress was included. This last treatment received four irrigations – besides the pre-planting irrigation; 22, 45, 65 and 80 days after planting (DAP). The treatment receiving water stress during the vegetative stage received no water at the 22 DAP irrigation; and the second water stress treatment received no water at 65 DAP. Data for soil moisture, nodule number and nodule dry weight were taken in 0.5 m^2 plots during several sampling dates.

The bean cultivars were harvested according to their physiological stage (R-8). Early maturing cultivars were harvested between 90 and 95 DAP and late maturing ones between 100 and 108 DAP. The harvested area was 0.80 m^2. Bean plants were separated into pods and straw, but millet was handled without separation of grain. All harvested materials were dried at 70°C, weighted and ground. Sub-samples were analysed for total N and ^{15}N isotope ratio by 1500 ANA N analyzer coupled to a VG mass spectrometer at the IAEA's Seibersdorf Laboratory. The proportion of N derived from atmosphere was calculated using the isotope dilution method (McAuliffe et al., 1958; Fried and Middelboe, 1977). Statistical analysis were done using the least significant different multiple-range test.

Results and discussion

Moisture stress treatments resulted in decreased soil moisture for nearly three weeks from 25 to 44 DAP and 60 to 79 DAP for each of the treatments at vegetative and reproductive stages, respectively (Table 1). Field capacity and perma-

Table 1. Soil moisture content during several sampling dates for the water stress treatments

Water stress treatment	Soil depth (cm)	Moisture content (%)									
		9[a]	21	22	40	44	47	73	79	83	100
Vegetative	0–30	39	34	33	25	24	46	50	42	50	39
Stage	30–60	37	32	33	32	30	40	41	37	43	36
Reproductive	0–30	39	35	45	34	32	46	28	25	49	38
Stage	30–60	36	33	39	35	32	38	31	29	42	36
Control	0–30	39	34	46	34	32	47	50	43	50	37
	30–60	37	33	38	36	33	40	40	36	40	35

[a] Days after planting.

nent wilting point of the soil was at 50 and 23%, respectively.

Water stress during vegetative growth significantly ($p < 0.05$) reduced the number of nodules in cultivars FM Bajio and Honduras-35 at the 42 DAP sampling date (Table 2). However, this did not occur in cv. Bayocel and Bat-477. Sinclair et al. (1988) and Kirda et al., (1989) reported that severe drought was required before decreases in nodulation occurred in soybean. Therefore, some cultivars may be more susceptible to the effect of water stress on nodulation than others. Similar trends were observed in nodule dry weight (Table 3). Nearly 100% reduction was observed for FM Bajio at this sampling date. This reduction was however only

Table 2. Nodule number during several sampling dates for the four dry bean cultivars under water stress treatments

Cultivar	Water stress	Number of nodules/m^2				
		36[a]	42	57	80	91
F.M. Bajío	Vegetative	328	192	1180	410	372
F.M. Bajío	Reproductive	279	324	582	268	60
F.M. Bajío	Control	299	454	558	624	127
Bayocel	Vegetative	282	346	1275	1128	1053
Bayocel	Reproductive	364	408	1023	592	340
Bayocel	Control	342	498	1113	868	639
Bat-477	Vegetative	396	357	1087	598	778
Bat-477	Reproductive	215	381	711	343	268
Bat-477	Control	221	360	1120	583	868
Honduras-35	Vegetative	313	360	882	877	429
Honduras-35	Reproductive	460	695	531	265	117
Honduras-35	Control	440	679	553	624	408

[a] Days after planting.

Table 3. Nodule dry weight during several sampling dates for the four dry bean cultivars under water stress treatments

Cultivar	Water stress	Nodule dry weight/m^2					
		36[a]	42	57	80	91	Average
F.M. Bajío	Vegetative	0.67	0.45	0.95	0.59	0.54	0.64
F.M. Bajío	Reproductive	0.77	0.88	0.41	0.33	0.10	0.49
F.M. Bajío	Control	0.43	0.87	0.70	0.73	0.22	0.59
Bayocel	Vegetative	0.53	1.08	1.73	2.60	2.13	1.61
Bayocel	Reproductive	0.63	1.16	1.83	0.81	0.52	0.99
Bayocel	Control	0.84	1.24	1.76	2.33	1.56	1.55
Bat-477	Vegetative	0.81	0.90	1.10	1.21	1.90	1.18
Bat-477	Reproductive	0.40	1.07	1.07	0.77	0.43	0.75
Bat-477	Control	0.38	1.01	1.79	1.29	1.87	1.27
Honduras	Vegetative	0.52	1.14	0.98	1.05	0.85	0.90
Honduras	Reproductive	0.75	1.25	0.85	0.30	0.09	0.65
Honduras	Control	0.65	1.64	1.00	0.52	0.81	0.92

[a] Days after planting.

temporary and after resumed water application nodule number and weight was even significantly ($p < 0.05$) larger than the control crop, which had not received any water stress. This was particularly important in cv. FM Bajio with number and nodule dry weight increasing 100 and 35% as compared to the control, twelve days after resumed water application. The rest of the cultivars did not show such fast recovery. Although Bat-477 is considered a water stress tolerant genotype (White, 1989), it recovered slowly after stress. With the exception of Bat-477, there was a clear tendency of the treatments which received water stress during the vegetative stage to keep their nodules for a longer period of time than the control. Nodulation was also significantly ($p < 0.05$) reduced during the last two samplings when water stress was imposed during the reproductive stage. However, this reduction was irreversible with no recovery after resumed water application. Bayocel and Bat-477 had significantly ($p < 0.05$) more nodules than the other cultivars at the last sampling date presumably due to their late maturity.

These results are in agreement with those reported by Kirda et al. (1989) in soybean, who found that earlier stress followed by later watering was less detrimental to nodule development than the reverse.

In general, water stress during the vegetative stage resulted in delayed maturity while water stress at the reproductive stage had the opposite effect (Table 4). Water stress prior to flowering could lead to the loss of the first blooms and therefore the competition for photosynthates

Table 4. Earliness to flowering and maturity of the four dry bean cultivars

Genotype	Water stress	Flowering days	Maturity
F.M. Bajío	Vegetative	48	97
F.M. Bajío	Reproductive	48	92
F.M. Bajío	Control	48	95
Bayocel	Vegetative	60	114
Bayocel	Reproductive	60	105
Bayocel	Control	60	109
Bat-477	Vegetative	55	110
Bat-477	Reproductive	54	105
Bat-477	Control	53	108
Honduras-35	Vegetative	54	98
Honduras-35	Reproductive	53	95
Honduras-35	Control	54	100

Table 5. Yield, nitrogen harvest and N_2 fixation

Genotype	Water stress	Grain yield (ton ha^{-1})	Harvest index	Nitrogen yield (kg ha^{-1})	Ndfa (%)	N_2 fixed (kg ha^{-1})
F.M. Bajío	Vegetative	2.89 bc*	0.57 ab	102 e	32	32
F.M. Bajío	Reproductive	2.23 d	0.55 ab	86 f	38	37
F.M. Bajío	Control	2.84 c	0.58 a	103 e	38	39
Bayocel	Vegetative	4.32 a	0.57 ab	155 b	64	99
Bayocel	Reproductive	3.03 b	0.51 bc	139 c	59	82
Bayocel	Control	4.35 a	0.55 ab	158 b	60	95
Bat-477	Vegetative	3.46 b	0.52 b	144 bc	59	85
Bat-477	Reproductive	2.33 cd	0.45 c	123 d	57	70
Bat-477	Control	4.23 a	0.59 a	179 a	58	104
Honduras-35	Vegetative	2.12 de	0.51 bc	83 fg	39	32
Honduras-35	Reproductive	1.54 e	0.50 bc	68 g	43	29
Honduras-35	Control	2.01 de	0.53 bc	92 ef	40	37

* Values followed by different letters are significantly different ($p < 0.05$).

between roots and pods might be delayed to the benefit of nodulation, as suggested by Hardy and Havelka (1976) and St. Claire et al. (1985).

Percentages of nitrogen derived from the atmosphere for cv Bayocel and Flor de Mayo were consistent with those obtained in previous experiments (Grajeda, 1990). It is also important to point out that Flor de Mayo and Honduras-35 are short season cultivars whereas Bayocel and Bat-477 are late maturity cultivars, with larger nitrogen fixing ability.

Even though water stress affected nodulation this was not translated into a significant decrease in percent nitrogen fixed (Table 5) which was contrary to previous studies (Kirda et al., 1989). One possible explanation for these results is that at the time of water stress during the reproductive stage the plants had already satisfied most of their nitrogen demand. According to Grajeda (1990), Bayocel and Flor de Mayo grown under similar conditions derived as much as 70% of the total N from the atmosphere 70 days after planting. Thus, the water stress at this advanced stage of growth seems not to effect N_2 fixation.

However, significant reductions ($p < 0.05$) in grain yields were observed when water stress was applied during reproductive stage. This observation is in agreement with most reports (Halterlein, 1989), and this was particularly noticeable in late maturity cultivars. In fact, harvest index of these cultivars was reduced significantly ($p < 0.05$) as compared to the early maturity cultivars (Table 5) as a result of drought. Late maturity plants need more time to remobilize carbohydrates to pods, and therefore they are not able to avoid or escape from stressing conditions.

References

Abdel-Ghaffar A S, Alter H A, El Halfawe M H and Abdel Salam A A 1982 Effects of inoculation, nitrogen fertilizer, salinity and water stress on symbiotic N_2 fixation by *Vicia faba* and *Phaseolus vulgaris* L. *In* BNF Technology for Tropical Agriculture. Eds. P H Graham and S C Harris. pp 153–159. CIAT, Cali, Colombia.

Durand J L, Sheehy J E and Michigan F R 1987 Nitrogenase activity, photosynthesis and nodule water potential in soya bean plants experiencing water deprivation. J. Exp. Bot. 38, 311–321.

Fried M and Middelboe V 1977 Measurement of the amount of nitrogen fixed by a legume crop. Plant and Soil 47, 713–715.

Grajeda-Cabrera, O A 1990 Cinetica de la fijaciçn de nitrogeno en frijol comun. Tesis de Mestria Centro de Invesgicacion y Estudios Avanzados del IPN. Unidad Irapuato, Gto. Mexico.

Halterlein A J 1988 Bean. *In* Crop Water Relations. Eds. I D Teare and M M Peet. pp 158–185. Wiley, New York.

Hardy R M F and Havelka U D 1976 Photosynthate as a major factor limiting nitrogen fixation with emphasis on soybeans. *In* Symbiotic Nitrogen Fixation in Plants. Ed. P S Nutman. Cambridge University.

Ibarra F, Pajarito A, Acosta-Gallegos J, Reyes E and Rodriguez R 1987 Frijol de temporal en Durango. *In* Avances de Investigacion Agricola en la region de los Llanos, Durango. CIFAP-Durango. pp 10–19.

Kirda C, Danso S K A and Zapata F 1989 Temporal water stress effects on nodulation, nitrogen accumulation and growth of soybean. Plant and Soil 120, 49–55.

McAuliffe C and Chamblee D S, Uribe-Arargo H and Woodhouse W W Jr 1958 Influence of inorganic nitrogen on nitrogen fixation by legumes as revealed by ^{15}N. Agron. J. 55, 334–337.

Pankurst C E and Sprent J I 1975 Effects of water stress on the respiratory and nitrogen fixing activity of soybean root nodules. J. Exp. Bot. 26, 287–304.

Saito S M, Nazareth M, Montanheiro S, Victoria R C and Reichardt K 1984 The effects of N fertilizer and growth of *Phaseolus vulgaris*. J. Agric. Sci. Camb. 103, 87–93.

Sinclair T R, Zimet A R and Muchow R C 1988 Changes in soybean nodule number and dry weight in response to drought. Field Crops Res. 18, 197–202.

Sprent J I 1971 The effects of water stress on nitrogen-fixing root nodules. New Phytol. 70, 9–17.

Sprent J I 1982 Nitrogen fixation in grain legumes in the UK. Phil. Trans. R. Soc. Lond. 296, 387–395.

St. Claire D A, Rosas J C, Bliss F A, Dubois J D and Burris R H 1985 Evaluation of N_2 fixation and N-partitioning in common beans using ^{15}N depleted $(NH_4)_2SO_4$ and acetylene reduction. Hortic. Sci. 20, 586.

White J M and Castillo J A 1988 Studies at CIAT on mechanisms of drought tolerance in common beans. Eds. J W White, G Hoogenboom, F Ibarra and S P Singh. pp 146–164. Docto. De Trabajop No. 41. CIAT, Colombia.

Zablotowicz R M, Focht D D and Cannell G H 1981 Nodulation and nitrogen fixation of field growth cowpeas as influenced by well irrigated and drought conditions. Agron J. 73, 9–12.

Plant and Soil **152**: 157–160, 1993.

PLSO BNF18

Utilizing the potential for increased nitrogen fixation in common bean

F.A. BLISS
Department of Pomology, University of California-Davis Davis, CA 95616, 8683, USA

Key words: common bean, *Phaseolus vulgaris* L.

Abstract

Many variables affect the amount of N_2 fixation that occurs in field-grown common beans (*Phaseolus vulgaris* L.). When environmental conditions are optimized, genetically superior plants that are nodulated with efficient rhizobia are able to fix enough N_2 to support grain yields of up to 2000 kg ha^{-1}. All of the required components are available for common bean to fulfill its potential as a nitrogen-fixing food crop. However, these components must be assembled and presented to growers in a usable package that is economically attractive.

The studies of N_2 fixation in common bean that have been supported by this project (see other chapters in this volume) reflect many of the types of production systems and sites common to the bean growing areas of Latin America. While it was not possible either to grow the entire range of cultivars that are preferred by local consumers or to duplicate the exact production practices of the farmers, there was no attempt to provide high levels of inputs to maximize yields in a way unattainable by farmers using limited purchased inputs. These results should reflect what can be expected regarding N_2 fixation if locally adapted cultivars are grown under conditions attainable by local farmers who are willing to incorporate some improved production practices such as inoculation, to ensure better response of superior bean cultivars.

The variable conditions under which common beans are grown contribute greatly to the differing amounts of plant N derived from N_2 fixation. The fertility of the soils relegated to bean production vary greatly in pH and major and minor elements, from inadequate amounts to levels toxic to plants and *Rhizobium*. High temperatures, especially in the upper layers of dry soils where bean roots normally proliferate if conditions are optimal, contribute greatly to lower N_2 fixation. Potentially high levels of N_2 fixation by lateral root nodules near the soil surface cannot be attained under moisture and temperature stress, and the resultant lack of N for the developing seeds is especially yield limiting. Small plants with limited growth, resulting in exposure of the soil to direct sunlight, contribute further to the high temperatures and dry conditions that are unfavourable to *Rhizobium* and lateral root growth. Protection of the soil surface should reduce moisture and temperature stress and enhance N_2 fixation.

Although native rhizobia usually are present in soils where beans are grown traditionally, the numbers of bacteria and the effectiveness of those present are often unknown. The procedures used to quantify rhizobial populations and to assess their effectiveness are usually not practical for the grower. When rhizobial inoculant is available, treatment of the seed or application to the soil is recommended to ensure the presence of effective *Rhizobium* strains.

Previous beliefs that crown nodules were of supreme importance and that active N_2 fixation dropped dramatically at the onset of seedfill have been shown to be only partially true. At least in improved bean genotypes, lateral root nodules account for a substantial portion of the total plant Ndfa (Wolyn et al., 1989) and active fixation occurs during much of the time of seed maturation (Attewell and Bliss, 1985). The important role of lateral root nodules should be recognized by rhizobiologists interested in assessing the competitiveness of *Rhizobium* strains. The disappointing field performance of strains found to be 'super' in the laboratory may not be completely due to their inability to compete with other strains. Inoculant placed on the seed or in the seed furrow is not likely to produce nodules on lateral roots because of the limited lateral mobility of *Rhizobium*. Competitiveness and the response of improved cultivars to inoculation with 'super' strains should be reassessed based on the current understanding of the importance of lateral root nodules to total plant N nutrition (Hardarson, 1989; Wolyn et al., 1989). When possible, alternative procedures of application should be studied to determine optimum inoculation practices for maximizing N_2 fixation in production fields.

The initial conclusions that common bean is a poor N_2 fixing legume were based on analyses of a few plant genotypes, often with the estimates being based on indirect measures such as the acetylene reduction assay taken at a few discrete times. When methods suitable for quantification are used properly, the amount of plant N expressed as either % Ndfa or total Ndfa has been found to be quite substantial for many bean lines grown under conditions reasonably favourable for growth and development. If the environment is too unfavourable for plant growth, surely the host and its symbiotic partner cannot be completely at fault! The farmer bears some responsibility for providing a decent production environment just as they would for more glamorous crops. To allow maximum expression of symbiotic potential other plant growth-limiting factors must also be corrected.

It has been common to observe locally-adapted bean cultivars able to fix at least 50 kg ha^{-1} and obtain 40–50% of the plant N from fixation. When these levels were attained and other factors were not greatly limiting, grain yields of from 1 to 1 1/2 T ha^{-1} were produced, as would be expected based on an N composition of grain and straw. While upon initial consideration, production of 50 kg N ha^{-1} from N_2 fixation may seem somewhat modest, this should be compared to recovering a similar amount *in the plant at harvest* from the addition of fertilizer N. In carefully-conducted grain legume experiments, rarely is more than 30–50% of the applied N recovered in the plant. If 50% (a liberal figure) recovery is assumed, application of 100 kg ha^{-1} of N must be applied to equal 50 kg ha^{-1} from N_2 fixation. Rarely, do expert extension specialists recommend application of more than 40–50 kg N ha^{-1} for optimum production. For beans grown on low-N soils (i.e., providing 20–40 kg N ha^{-1} derived from the soil), addition of 40–50 kg N ha^{-1}, of which less than 50% will be recovered in the plant, is not adequate to produce the high yields of 2 T ha^{-1} or more that agronomists talk fondly about. To reach those yields, the soils must be inherently N fertile, substantial amounts of fertilizer must be added in small increments throughout the growing season, or the bean plant must fix amounts of from 50 to 100 kg N ha^{-1}. The first two conditions are often difficult for the low-input farmer to attain realistically. The latter is attainable more readily if superior cultivars and improved production practices are employed.

The simplest, most straight-forward approach to breeding for improved N_2 fixation is to identify well-adapted local landrace cultivars with the capacity for high N_2 fixation. In most of the trials there appeared to be plants with this ability. However, promising lines may not have all the other required traits, e.g., disease resistance, earliness, satisfactory yield, etc. If not, the breeder can choose appropriate breeding methods to transfer those other traits into the high N_2 fixing local cultivars.

Just as likely, the currently-recommended bean cultivars may not have the potential for high N_2 fixation. If not, the first step is to formulate improvement objectives. What are the desired seed yield, the total plant N required from N_2 fixation to reach that yield, and the % Ndfa to replace the N removed in the seed harvest? For the cultivar to be nitrogen sustainable, the plant % Ndfa must be equal to the N Harvest Index at the prescribed seed yield. High performance expectations will not be easily attainable. Therefore, the next step is to identify superior sources of plant genetic variabilty for enhanced N_2 fixation.

Studies from several programmes during the

past 15 years, including those reported in this project, have identified genetically high N_2-fixation lines. Superior parents appear to be reasonably plentiful, although often they do not have other traits one may desire, e.g., preferred seed traits, growth habit, earliness, etc. The transfer of high N_2 fixation ability should be successful when a method that relies on the production of families is used, i.e., modified pedigree selection, inbred backcross line method. The populations of segregating families should be grown at low-N field sites, where it is possible to identify superior families based on replicated trials. Methods that allow accurate estimates of N_2 fixation, e.g., ^{15}N-isotopic methods, N difference method, total seed N, etc., and facilitate discrimination between superior and inferior lines are essential. The family rather than a single plant is preferred as the unit of selection, since often the estimation method is plant destructive. Also, some of the plants in the family experimental unit can be used to estimate N_2 fixation while the remainder are used to evaluate other traits, e.g., yield, seed traits, etc. The inclusion of a suitable non-fixing reference crop is essential for some methods of estimation and it is a valuable visual indicator of the soil N level in all field trials.

The identification of superior landrace cultivars and the production of high N_2-fixing lines through breeding must be followed by field evaluations to determine their relative performance compared to the accepted cultivars that are commonly grown. If the experimental lines purported to have high N_2 fixation are to be evaluated fairly for this trait, the regional yield trials must be conducted on low-N soils with application of no more than a minimum amount of N at planting time. Ideally, the trials should reflect the conditions experienced routinely on farms where low amounts of purchased inputs are utilized. This is an excellent opportunity to evaluate new lines in 'on-farm' trials, where seeds are given to the grower who is asked to plant them in the production field. Evaluation then becomes a joint venture of both grower and researcher in a setting familiar to the user. The degree of success or failure in a real-world setting will become apparent more quickly.

Often, it is difficult to fairly evaluate potentially high N_2-fixing experimental bean lines in the yield trials conducted on experiment stations. Usually, fertility and management conditions are optimized and the ability of the experimental lines to excel because of increased N_2 fixation is obscured. Commonly, a yield trial is said to be 'good' if the Coefficient of Variation (CV) is low. Of course, one way to achieve a low CV is to increase the overall trial mean yield – through the provision of maximum inputs – even though this may be entirely incompatible with the needs of the farmer. A better way to achieve a lower CV in a bean yield trial receiving low N fertilizer is by reducing the standard error of the mean through additional replication. In this way, promising lines able to fix N_2 effectively will not be discarded for the wrong reason. The objectives of improvement programmes must be kept clearly in focus. If selecting for increased N_2 fixation was an important initial objective, final evaluation for release must include this important trait and should not reflect only disease resistance, earliness, high yield under maximum input, etc., unless those traits truely are of singular importance.

Breeders working on other characters unrelated to nitrogen fixation should preferably increase their breeding lines with *Rhizobium* and not apply any N fertilizer. By doing so they would indirectly produce lines with increased nitrogen fixation.

All of the required components are available for common bean to fulfill its potential as a nitrogen-fixing food crop. However, in most cases, they are not all in place in a single useable package. What remains is for the agronomist, breeder, microbiologist, farmer and others to combine the essential components – adapted high-fixing cultivars, effective rhizobial inoculant, practical agronomic practices, and economic management – into a package that the farmer has helped to assemble and will use for optimum production.

References

Attewell J and Bliss F A 1985 Host plant characteristics of common bean lines selected using indirect measures of N_2 fixation. *In* Nitrogen Fixation Research Progress: Proc. of the 6th Intern. Sympos. on Nitrogen Fixation. Eds. H J Evans, P J Bottomley and W E Newton. pp 3–9. Martinus Nijhoff Publishers, Dordrecht, The Netherlands.

Wolyn D J, Attewell J, Ludden P W and Bliss F A 1989 Indirect measures of N_2 fixation in common bean (*Phaseolus vulgaris* L.) under field conditions; The role of lateral root nodules. Plant and Soil 113, 181–187.

Developments in Plant and Soil Sciences

1. J. Monteith and C. Webb (eds.): *Soil Water and Nitrogen in Mediterranean-type Environments*. 1981 ISBN 90-247-2406-6
2. J. C. Brogan (ed.): *Nitrogen Losses and Surface Run-off from Landspreading of Manures*. 1981 ISBN 90-247-2471-6
3. J. D. Bewley (ed.): *Nitrogen and Carbon Metabolism*. 1981 ISBN 90-247-2472-4
4. R. Brouwer, I. Gašparíková, J. Kolek and B. C. Loughman (eds.): *Structure and Function of Plant Roots*. 1981 ISBN 90-247-2510-0
5. Y. R. Dommergues and H. G. Diem (eds.): *Microbiology of Tropical Soils and Plant Productivity*. 1982 ISBN 90-247-2624-7
6. G. P. Robertson, R. Herrara and T. Rosswall (eds.): *Nitrogen Cycling in Ecosystems of Latin America and the Caribbean*. 1982 ISBN 90-247-2719-7
7. D. Atkinson, K. K. S. Bhat, M. P. Coutts, P. A. Mason and D. J. Read (eds.): *Tree Root Systems and Their Mycorrhizas*. 1983 ISBN 90-247-2821-5
8. M. R. Sarić and B. C. Loughman (eds.): *Genetic Aspects of Plant Nutrition*. 1983 ISBN 90-247-2822-3
9. J. R. Freney and J. R. Simpson (eds.): *Gaseous Loss of Nitrogen from Plant-Soil Systems*. 1983 ISBN 90-247-2820-7
10. United Nations Economic Commission for Europe (ed.): *Efficient Use of Fertilizers in Agriculture*. 1983 ISBN 90-247-2866-5
11. J. Tinsley and J. F. Darbyshire (eds.): *Biological Processes and Soil Fertility*. 1984 ISBN 90-247-2902-5
12. A. D. L. Akkermans, D. Baker, K. Huss-Danell and J. D. Tjepkema (eds.): Frankia *Symbioses*. 1984 ISBN 90-247-2967-X
13. W. S. Silver and E. C. Schröder (eds.): *Practical Application of* Azolla *for Rice Production*. 1984 ISBN 90-247-3068-6
14. P. G. L. Vlek (ed.): *Micronutrients in Tropical Food Crop Production*. 1985 ISBN 90-247-3085-6
15. T. P. Hignett (ed.): *Fertilizer Manual*. 1985 ISBN 90-247-3122-4
16. D. Vaughan and R. E. Malcolm (eds.): *Soil Organic Matter and Biological Activity*. 1985 ISBN 90-247-3154-2
17. D. Pasternak and A. San Pietro (eds.): *Biosalinity in Action*. Bioproduction with Saline Water. 1985 ISBN 90-247-3159-3
18. M. Lalonde, C. Camiré and J. O. Dawson (eds.): Frankia *and Actinorhizal Plants*. 1985 ISBN 90-247-3214-X
19. H. Lambers, J. J. Neeteson and I. Stulen (eds.): *Fundamental, Ecological and Agricultural Aspects of Nitrogen Metabolism in Higher Plants*. 1986 ISBN 90-247-3258-1
20. M. B. Jackson (ed.): *New Root Formation in Plants and Cuttings*. 1986 ISBN 90-247-3260-3
21. F. A. Skinner and P. Uomala (eds.): *Nitrogen Fixation with Non-Legumes* (Proceedings of the 3rd Symposium, Helsinki, 1984). 1986 ISBN 90-247-3283-2
22. A. Alexander (ed.): *Foliar Fertilization*. 1986 ISBN 90-247-3288-3
23. H. G. v.d. Meer, J. C. Ryden and G. C. Ennik (eds.): *Nitrogen Fluxes in Intensive Grassland Systems*. 1986 ISBN 90-247-3309-X
24. A. U. Mokwunye and P. L. G. Vlek (eds.): *Management of Nitrogen and Phosphorus Fertilizers in Sub-Saharan Africa*. 1986 ISBN 90-247-3312-X
25. Y. Chen and Y. Avnimelech (eds.): *The Role of Organic Matter in Modern Agriculture*. 1986 ISBN 90-247-3360-X
26. S. K. De Datta and W. H. Patrick Jr. (eds.): *Nitrogen Economy of Flooded Rice Soils*. 1986 ISBN 90-247-3361-8
27. W. H. Gabelman and B. C. Loughman (eds.): *Genetic Aspects of Plant Mineral Nutrition*. 1987 ISBN 90-247-3494-0
28. A. van Diest (ed.): *Plant and Soil: Interfaces and Interactions*. 1987 ISBN 90-247-3535-1

Developments in Plant and Soil Sciences

29. United Nations Economic Commission for Europe and FAO (eds.): *The Utilization of Secondary and Trace Elements in Agriculture.* 1987 ISBN 90-247-3546-7
30. H. G. v.d. Meer, R. J. Unwin, T. A. van Dijk and G. C. Ennik (eds.): *Animal Manure on Grassland and Fodder Crops.* Fertilizer or Waste? 1987 ISBN 90-247-3568-8
31. N. J. Barrow: *Reactions with Variable-Charge Soils.* 1987 ISBN 90-247-3589-0
32. D. P. Beck and L. A. Materon (eds.): *Nitrogen Fixation by Legumes in Mediterranean Agriculture.* 1988 ISBN 90-247-3624-2
33. R. D. Graham, R. J. Hannam and N. C. Uren (eds.): *Manganese in Soils and Plants.* 1988 ISBN 90-247-3758-3
34. J. G. Torrey and J. L. Winship (eds.): *Applications of Continuous and Steady-State Methods to Root Biology.* 1989 ISBN 0-7923-0024-6
35. F. A. Skinner, R. M. Boddey and I. Fendrik (eds.): *Nitrogen Fixation with Non-Legumes* (Proceedings of the 4th Symposium, Rio de Janeiro, 1987). 1989 ISBN 0-7923-0059-9
36. B. C. Loughman, O. Gašparíková and J. Kolek (eds.): *Structural and Functional Aspects of Transport in Roots.* 1989 ISBN 0-7923-0060-2; Pb 0-7923-0061-0
37. P. Plancquaert and R. Haggar (eds.): *Legumes in Farming Systems.* 1990 ISBN 0-7923-0134-X
38. A. E. Osman, M. M. Ibrahim and M. A. Jones (eds.): *The Role of Legumes in the Farming Systems of the Mediterranean Areas.* 1990 ISBN 0-7923-0419-5
39. M. Clarholm and L. Bergström (eds.): *Ecology of Arable Land – Perspectives and Challenges.* 1989 ISBN 0-7923-0424-1
40. J. Vos, C. D. van Loon and G. J. Bollen (eds.): *Effects of Crop Rotation on Potato Production in the Temperate Zones.* 1989 ISBN 0-7923-0495-0
41. M. L. van Beusichem (ed.): *Plant Nutrition – Physiology and Applications.* 1990 ISBN 0-7923-0740-2
42. N. El Bassam, M. Dambroth and B.C. Loughman (eds.): *Genetic Aspects of Plant Mineral Nutrition.* 1990 ISBN 0-7923-0785-2
43. Y. Chen and Y. Hadar (eds.): *Iron Nutrition and Interactions in Plants.* 1991 ISBN 0-7923-1095-0
44. J. J. R. Groot, P. de Willigen and E. L. J. Verberne (eds.): *Nitrogen Turnover in the Soil-Crop System.* 1991 ISBN 0-7923-1107-8
45. R. J. Wright, V.C. Baligar and R. P. Murrmann (eds.): *Plant-Soil Interactions at Low pH.* 1991 ISBN 0-7923-1105-1
46. J. Kolek and V. Kozinka (eds.): *Physiology of the Plant Root System.* 1992 ISBN 0-7923-1205-8
47. A. U. Mokwunye (ed.): *Alleviating Soil Fertility Constraints to Increased Crop Production in West Africa.* 1991 ISBN 0-7923-1221-X; Pb 0-7923-1222-8
48. M. Polsinelli, R. Materassi and M. Vincenzini (eds.): *Nitrogen Fixation* (Proceedings of the 5th Symposium, Florence, 1990). 1991 ISBN 0-7923-1410-7
49. J. K. Ladha, T. George and B. B. Bohlool (eds.): *Biological Nitrogen Fixation for Sustainable Agriculture.* 1992 ISBN 0-7923-1774-2
50. P. J. Randall, E. Delhaze, R. A. Richards and R. Munns (eds.): *Genetic Aspects of Plant Mineral Nutrition.* 1993 ISBN 0-7923-2118-9
51. K. S. Kumarasinghe and D. L. Eskew (eds.): *Isotopic Studies of Azolla and Nitrogen Fertilization of Rice.* 1993 ISBN 0-7923-2274-6
52. F. A. Bliss and G. Hardarson (eds.): *Enhancement of Biological Nitrogen Fixation of Common Baen in Latin America.* 1993. ISBN 0-7923-2451-X

Kluwer Academic Publishers – Dordrecht / Boston / London